The Drone Book

2018 Edition

A little bit about everything you need to know about drones

Chris Laforêt

The Drone Book
A little bit about everything you need to know about drones
2018 (2nd) Edition

ISBN: 098926534
ISBN-13: 978-0-9892653-4-8

Cover Design and Photography: Chris Laforêt
Cover Fonts: Trajan Pro and Gautami
Illustrator: Lindsay Thompson
Photographers: Chris Laforêt, Bekah Laforêt

 Like us on **Facebook**
www.facebook.com/EverythingAboutDrones

This book is dedicated to all the men and women whose brilliant ideas and inventions have opened the doors of flight to the masses.

"If you can walk away from a landing, it's a good landing. If you use the airplane the next day, it's an outstanding landing."

Chuck Yeager

Joshua 24:15b

Contents

Introduction

Drones are exciting and fun. Even so there are seemingly unlimited types and brands to choose from so choosing one is not necessarily an easy task. In addition to this, drones introduce a need to learn new tricks so that they can be flown safely. Then there is an entire landscape of regulation that involves drones since they operate in regulated airspace.

It is the purpose of this book to be a "go-to" reference about all things drone-related. The Internet has many excellent sources of information but it also has a number of sites filled with misinformation. When I first started flying drones at the beginning of 2014, I tripped into a number of pitfalls because of some of these misleading sites and I wished that there was a single book that could reside on my bookshelf (real or virtual) which could outline the basics for me. Such a book would be a virtual North-star to help me navigate the complexities of the Dronespace.

So, here is my attempt to provide such a reference. It is my hope that this will become one of those books which helps you grow your drone knowledge. It is my hope that it will be updated frequently enough to keep up with this fast-growing industry and will continue being that "go-to" for years to come.

My experience with drones is limited at this time to rotorcraft. Rotorcraft are drones which use multiple propellers to achieve lift (like a helicopter) and to maneuver. The most common type of rotorcraft drone is the Quadcopter design which is characterized by 4 propellers pointing upwards. There are also Hexacopters (6 props) and Octacopters (8 props). Another type of drone is a fixed-wing drone similar to an airplane.   There are also some newly created tethered drones which have controllers built into their tether handles. I might mention either of these occasionally in the book but since I don't have experience with them currently, I cannot expound about how they fly. Maybe in the next while I will get my hands on one and a future release of this book will delve into them.

Hopefully you will find this book useful. It is almost impossible to keep on top of every aspect of the Dronespace in any one edition of a book, but it is my goal to keep updating it to keep it relevant over time. I am always eager to hear suggestions on how future editions of the book could be improved and any feedback that you can provide. You can find the book's page on Facebook (www.facebook.com/EverythingAboutDrones) and send feedback there. Don't forget to Like the page and recommend the page to your friends!

Finally, I would like to take the time to thank those who helped me pull this book together. I would like to thank to Lindsay "L.T" Thompson, an accomplished filmmaker in her own right, for her work on the illustrations. A warm thank-you to those who proofed the book and made suggestions to improve it: Jorge Vazquez, Forris "Butch" Day, Joe Valasquez from DroneScape and organizer of the Charlotte Drones meetup group, and Rick Simmons III. I would be remiss in not thanking my wife, Sherry, for encouraging me to buy my first drone and our daughter, Bekah, for her photography used in this book.

Welcome to the Dronespace

Congratulations. Since you are reading this book, you either have purchased a drone or are planning to in the near future. Drones are so much fun and have tremendous practical applications that many who watch the industry predict that it will grow explosively over the next decade. This book exists to be a go-to reference for as many things to do with drones as possible. While some aspects of the Dronespace may change slightly, the basics will not change that drastically now that the FAA has released its Part 107 rules. Also, while there will be some innovative companies that choose to use drones in unforeseen ways, there are many applications that already exist and will be the basis for other innovations.

What is the Dronespace? For the purposes of this book, it is the airspace which is occupied by one or more sUASes (small Unmanned Aerial System or "drones") at a single point in time. As you will discover, the Dronespace can be the airspace above a beautiful mountain, over fields on a plain, over your own back yard, or even inside a large building such as an arena. Once a drone is active and in the air, Dronespace exists.

Some Dronespace is regulated by one or more agencies, at least here in the United States. In every one of the examples above except for inside buildings, the FAA potentially has jurisdiction over the space that drone is occupying. I say "potentially" because there is at least one condition under which the FAA has released its dominion for now, namely if that drone weighs less than 0.55 pounds when fully laden. On a drone that exceeds this weight, the FAA is in full control and there may be other agencies that also control that airspace. For example, the airspace above any asset owned by the National Park Service is off-limits for any drones even if it is approved by the FAA for flight. The same is true for a growing number of states' park systems. Additionally, there are blocks of airspace that are restricted by the military and homeland-security to all flights.

When larger drones first emerged on the market, the skies were wide open and rules were few and far between. Now the opposite situation exists. This does not mean that you and I cannot enjoy flying our drones and using them to capture great video and photographs, but it does mean that we need to be aware of what we are doing. Which is where this book comes in. There are a large number of questions that surface online and many of them have differing answers which are just confusing. Hopefully this book will cover the basics of what you need to know and will help you make the most of what you read online.

The Good, the Bad, and the Ugly

Drone equipment has emerged over the past few years and people have become excited with all of its new technology. Sales of drones worldwide are skyrocketing and the predictions for growth of this market are very positive. The FAA has estimated that the sales of drones in the US will increase from 2.5 million in 2016 up to 7 million in 2020! It is certainly a booming market with no end in sight.

The hot topic at the latter half of 2016 is the large number of people who have registered to take the FAA remote pilot airman licensing exam. Now that the FAA has released 14 CFR Part 107 (referred to as simply "Part 107") regulations, there is a clear path for commercial drone operations. Before this, the FAA had an arbitrary application process for what was called a "333 exemption" to grant drone pilots limited permission to use drones commercially. It was slow and cumbersome. Many of us could not wait for August 29th, 2016 to roll around so that we could take the test and become officially licensed.

Prior to the end of December 2015, there were no requirements to register drones in the US. The FAA introduced a program for registration that went into effect just before the holiday season that year. As of December 21, 2015 all drones that weighed 0.55 pounds through just less than 55 pounds at takeoff had to be legally registered. The registration "tail number" is required to be marked clearly on the drone.

All of these regulations came into effect as the FAA worked with the drone community, pilots' organizations, and other lawmakers to "reign in the madness" that had started making news. There were incidents that received considerable press here in the US and abroad of drone users flying their drones in dangerous ways. Pilots on passenger aircraft on approach and takeoff at many of the nation's airports reported near-misses with drones that were being flown in restricted flight paths. Irresponsible drone users hampered rescue and fire-fighting operations by overflying disaster areas which caused official aircraft to be grounded until the threats could be contained. Drone pilots flew their craft over the White House and the Japanese Prime Minister's residence which sparked security concerns.

In some cases, instead of merely trespassing in airspace, some drone users actually caused physical and legal damage. There were cases of drones crashing into cars on busy city streets, voyeurism claims, and injuring people at sports events. Also, there were documented cases where drones were used to deliberately pester wildlife. Each of these events by irresponsible individuals helped put a black eye on what can be a lucrative and fun sport.

On the other side of the coin, there are so many positive applications for drone technology. Police and Fire departments have started using drones for search-and-rescue and fire-fighting visualization. Farmers use drones to save time doing inspections of their fence-lines and crops and are able to react to issues quicker. Cell and TV tower operating companies started using drones to do inspections of tall towers and are able to minimize putting a person's life at risk doing a climb unless absolutely necessary.

It is important that we, the citizens of the Dronespace, understand the laws of the land and apply them. It is also critical that each of us express our craft carefully and non-dramatically so that we can keep the bad spotlight off of this sport. Personally, every time I take off, I say to myself, "Don't be the Bozo who does something stupid with a drone!" It would suck to be quoted in law journals for years to come as being the person who accidentally brought down an airliner or caused a multiple-car accident!

Introduction to the 2018 (2nd) Edition

Drones are still exciting and fun and the Dronespace continues growing by leaps and bounds. It has been slightly less than a year since the first release of this book and there have been many new drones released, some changes in drone laws, new software capabilities have come out, and more practical uses for drones have been experimented with.

We have seen the merger of 3DR with DJI during the year. DJI has been such a drone powerhouse that 3DR. Unable to beat the competition, they have joined them by incorporating DJI support into their software. GoPro entered the drone market at the end of 2016 with its very promising and much-touted Karma, but technical issues forced them to make several painful recalls. They remain committed to their corporate course and re-released the Karmas after fixing the issues. DJI followed their very successful Mavic small drone with the even smaller Spark and added new gesture technology recognition to permit the pilot to signal some basic functions with hand gestures!

As I write this, commercial developers such as PrecisionHawk are perfecting applications that integrate tracked drone activity with traditional manned air traffic on ATC consoles. As fast as this technology is growing, I am sure this will be the main topic of interest in the next edition of the book!

Also included with this edition is a slew of new information concerning fixed-wing drone flight. This parallels my own personal development and probably would appeal to serious licensed drone pilots who desire to move into survey and agricultural drone work.

Throughout the year in 2017, I attended a number of conferences and worked personally with people who have started using drones in the fire and search-and-rescue realms. Drones have proven themselves useful in helping such efforts after natural disasters and to supplement the efforts of units on the ground.

I have also spoken with drone experts who have done joint exercises with the military on using drones to speed up location and rescue operations. The simple drone can get off the ground quicker and, using FLIR and other imaging techniques, hone in on a missing and/or injured person's location before the rescue helicopter can even get off the ground. This leads to quicker recovery of the person by the rescue crew, something that hopefully would improve that person's chances of survival. While writing this edition, the news of a drone fitted with infrared technology which located a missing woman in Shelby County, Indiana after the K-9 units lost her scent. The drone found her within 5 minutes and its location served as a mark for rescuers to reach her. Other stories of drones being used for successful S&R operations emerged from the Houston, TX area in the aftermath of Hurricane Harvey. As the technology improves and miniaturizes, I can see a drone in every rescue vehicle as a valuable tool in the arsenal of first-responders.

Needless to say, 2018 is going to be another very exciting year for drone development and use as new technologies and regulations come into the Dronespace. Drones, while still fun hobby "toys," are beginning to really flex their muscles as serious commercial vehicles. All we can do is to sit back and enjoy what is going to happen next!

Using Drones in the USA

The Benefits of Flying Legally

Let's admit it, there is a little bit of "bad-boy" or "bad-girl" in all of us. Why be legal? Why obey the laws of the land, or the air? This attitude is understandable and in terms of the wide-open landscape of the Dronespace, it seems reasonable to be an outlaw, doesn't it?

Of course, if we think about it a bit longer, we also may realize that most of us tend to be moderately law-abiding. While we are driving down the Interstate, we may be tempted to add a few miles-per-hour to the posted limit, but few of us dare to drive at racing-car speeds on public roads. We tend to consider that there are consequences to our actions and sometimes these consequences can be fatal. This internal mechanism tends to keep us all driving and behaving reasonably.

A drone may seem small and incapable of causing real problems but if we realize that if birds can cause aircraft to crash, then it is possible that a drone can too. Another thought that should make us pause and consider the danger of a drone is that if a brick falling through the air can kill someone, perhaps a drone can do the same thing. There are many ways that drones could be dangerous to life and limb.

There are other possibly legal ramifications to drone use that reasonable people would agree with. Some of these involve invasions of privacy. Others involve danger to pets and to wildlife. Yet others have to do with liability issues and damage to property.

So, why should flying legally be something we care about? First of all, it makes us good, responsible citizens in the Dronespace. It really isn't cool to be known as the person who caused new laws to be passed that take away the fun and profit that drones bring to countless others. Second, as the activity of drone flying becomes more popular, there may be a requirement to carry insurance or your drone won't turn on (actually, I advocate having coverage either way later in this book). Insurance companies will use whatever statistics they can to minimize their monetary risk so if you are on-record as flying illegally and/or recklessly you may end up in a far more expensive bracket or, worse, be uninsurable! Finally, drone flight can be mentally draining at times. Wouldn't it be good to not have to be concerned about being illegal while flying? That is one less problem to worry about.

The Role of the FAA in Dronespace

In the United States, anything that flies is technically under the jurisdiction of the Federal Aviation Administration (FAA). This agency regulates the airspace above all US territory. A drone being flown outdoors, by definition, is under the FAA's jurisdiction regardless of if it is being flown over land or territorial waters. Even if the drone is being flown on your personally owned property, the airspace is not considered yours but the FAA's.

However, if you fly your drone within the airspace of a closed structure such as an arena or a warehouse, the FAA does not regulate that activity. Likewise, the FAA has partially relinquished jurisdiction on regulating toy drones that weigh less than 0.55 pounds on takeoff. These small drones are not required to be registered or regulated otherwise.

The FAA's role begins with drones that weigh 0.55 pounds upwards. The new Part 107 regulations are specific to sUAS (Small Unmanned Airborne System), or drones, from 0.55 pounds fully loaded (laden) weight at takeoff up to 55 pounds laden weight at takeoff. This is the type of drone that this book covers. Anything 55 pounds or more would be handled by another set of FAA regulations. Anyway, sUASes that fall in the scope of Part 107 are either for hobbyist/enthusiast ("personal") use or for commercial use.

Personal use does not mean that drones can be flown however one wants. They are still under the regulations for airspace which limit the times of day they can be flown and where they can be legally flown. Also, personal use drones are required to be registered and a registration number attached to, or written on, the drone itself. What defines the difference between personal use and commercial use is if the drone will be used in any way for profit or to advance a business (such as for visual inspection of structures or roofs or taking real estate photos for a property for sale). The FAA has a page dedicated to personal use of drones www.faa.gov/uas/getting_started/fly_for_fun/ that is worth looking over.

As soon as someone who has a drone decides to use that drone for a commercial endeavor, Part 107 prescribes a different path. The drone is registered for a tail number as a commercial drone and the commercial operations need to be conducted with an FAA licensed Pilot-in-Charge (PIC) overseeing the activity. Insurance companies also, like the FAA, make a clear delineation between non-commercial and commercial drone use. We will discuss this in a later chapter.

It is important to understand that the FAA has defined specific no-fly zones around airports, limited the height of operations to 400 feet above ground level, restricted flight over stadiums and emergency response areas, and require the drone to be within sight of the operator at all times. These are among the requirements that exist for both personal and commercial drone use. We can be sure that the FAA can and will prosecute egregious violations of the rules to bring the Dronespace into some semblance of order to maintain safety for the countless aircraft that occupy the same airspace.

The Challenge of Living in Shared Airspace

If the previous section seemed worrisome, my purpose was not to be overly scary about the FAA's role. What we all need to grasp is that the FAA has to guard the transit of all aircraft in the airspace over our heads. It is important that we understand that with one foolish move, we may cause a major accident that can result in a large loss of life. This is because the Dronespace overlaps that airspace and drone use has a number of inherent limitations that affect the normal users of that airspace.

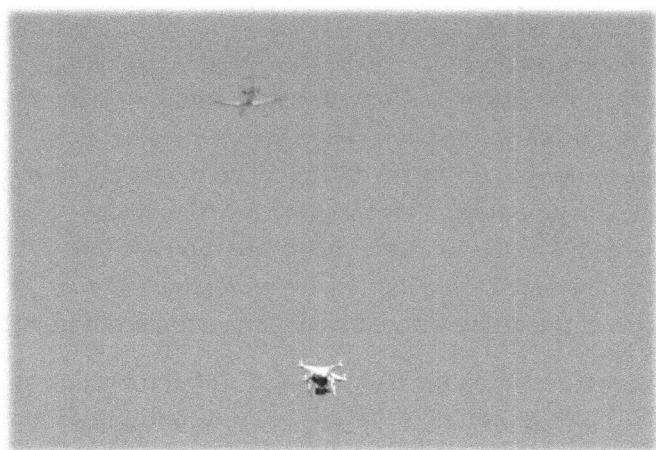

Figure 1 Drones fly in shared airspace

Drones do not have transponders so they will not show up on Air Traffic Control (ATC) or aircraft screens. They are small and easy to miss from the limited view of an airplane. Finally, most drone operators do not use radios tuned to local airport UNICOM/CTAF frequencies so that they can announce their activity and few contact airport management or ATC to let them know where they are operating. All of these limitations make it hazardous to regular air traffic which is used to the other occupants of the airspace being visible and making their locations and intentions known.

With these challenges in mind, the FAA's rules exist to separate the Dronespace from the rest of airspace as much as possible. Their rules concerning not flying within 5 miles of an airport (a minimum general guideline), staying below 400 feet above ground level, keeping the drone in line of sight, and not flying at night or when visibility is poor all are directly aligned with the safety of air traffic. The challenge each of us has to face is are we going to respect those rules or are we going to live on the ragged edge and possibly go down in history as the first drone user to cause a major plane crash!

Studying for the Remote Pilot Exam

Many of the readers of this book, I would imagine, are interested in exploiting their drones for profit. You may be a filmmaker, or real-estate agent, or farmer, or someone who wants to use a drone for inspection and imagery. Either way, you will need to think about having your drone registered for commercial use with the FAA and also ensure that your commercial ventures are overseen by a licensed Remote Pilot-in-Charge (PIC). Do be aware that even if you are the owner of the drone, you do not have to be the PIC. Likewise, you may be the operator handling the controls of the drone without having to be the licensed PIC.

A PIC is in charge of planning and executing the flight using all of the tools that all pilots have at their disposal. The requirement is that the PIC is responsible for all aspects of the flight, from the drone's maintenance and airworthiness, its loading and preparation for flight, all the way through the execution of the flight itself. The PIC must be either able to see the drone or be connected by radio to a Visual Observer (VO) who can relay the information about the drone's position back to the PIC. This requirement keeps the drone always in line of sight. Finally, the PIC must be able to immediately take over the controls of the drone should the need arise.

The FAA's requirements for licensing a person as a remote PIC involve taking a knowledge exam at a licensed FAA testing center. Someone who is already a pilot can take a different test and add remote PIC to their existing license. This exam includes information concerning weather, airports, planning, notification, and other items that are common to most pilot ground-school coursework. There are a growing number of private entities that permit taking this coursework over a day or two, many times over a weekend.

I personally took advantage of one of these programs (offered by DartDrones) and appreciated how it divided up the coursework into manageable pieces. Being taught by a pilot helped a great deal because he was able to challenge everyone in the class to think through scenarios. Much of the FAA's exam is tricky in the sense that you need to apply knowledge and reason to arrive at the right answer. In this situation, it is easy to second-guess yourself so it is worthwhile to gain as much experience as possible in answering such questions and being sure of your answer.

I recommend following that same route but it is not always feasible for everyone. There is a lot of information online that can be used to self-study. Some of the material is put out by the FAA itself and there are some reputable sites run by pilots which definitely serve to drill the information into one's brain. Even if you do decide to go to a ground-school, these sites still are helpful for providing greater understanding of specific subjects. The FAA has a sample exam available online and many sites also have sample exams and questions to help you to prepare.

The exam itself costs about $150 to take. If you do not pass it, you will be able to retake the test after 14 days. The following link will provide resources for anyone interested in becoming licensed as a remote PIC, www.faa.gov/uas/getting_started/fly_for_work_business/becoming_a_pilot/. The FAA also has a number of links for anyone interested in sUAS operation (personal or commercial) and the following FAQ page is a good springboard to use www.faa.gov/uas/faqs/.

Remember to Check State and Local Ordinances

Although the FAA maintains control over the airspace in the US, and technically individual states do not have such control, in terms of the Dronespace this is not the case. When drones emerged on the market, the FAA did not have any specific laws targeting these new devices and years went by until the lawmaking processes caught up with technology. Some states and agencies took the initiative and enacted laws and rules to rein in the wilder problems that occurred in Dronespace.

It is important to check on the laws in your state. For example, I am located in North Carolina and the NCDOT has its own additional examination procedure online to approve drone operators. There are a number of specific laws in North Carolina that are specific to drones and which address trespassing issues, voyeurism laws, wildlife protection, and use over state-owned parks. There are penalties that are attached to infractions of these specific laws which would present a hardship to many a drone hobbyist.

The National Park Service also became concerned about the non-regulated use of drones and they introduced (and still maintain) a strict no-fly zone over all of their properties. Some cities and townships also have ordinances against drone use, violation of which could result in penalties and/or confiscation of one's drone.

North Carolina's state-owned parks and historic sites also are No-Fly Zones now with a policy similar to the National Park Service's. However, unlike the NPS, permission can be sought to fly and can be granted by the park's manager. Other states and municipalities may have similar rules on their parks. It is encumbent upon the drone operator to check these out.

The FAA has efforts underway to create a standardized process for marking and tracking "No Drone Zones." Until this is in full use, we users of the Dronespace need to be as informed as possible and try to reflect positively on our sport. Sadly, even in 2017, even the NPS has not been diligent in posting these.

Do You Need Insurance?

Insurance is either a colossal waste of money or a tremendous benefit depending if you actually end up needing it. As we approach this subject, consider that insurance means two distinct things and we need to weigh each of these carefully in order to make a decent decision.

The first kind of insurance is replacement insurance, something that we may purchase to cover our lovely drone should it accidentally crash into a tree or crater itself into the ground. Essentially, choosing this insurance coverage depends upon the cost of your drone and its attached equipment, how sure you are that you (and other users of it) can be a safe, and how much you are willing to pay out of pocket to repair or replace it. Drone coverage is expensive since it is an emerging market short on safety statistics and what statistics do exist are abysmal!

The second kind of insurance is liability insurance and it serves to cover property that may be damaged and people that may be injured as a direct result of your drone use. These expenses can well outstrip the cost of a new drone by a factor of 10, 100, or even 1000 times! If you are uninsured you run the risk of losing everything if your drone crashes into something or someone. Bear in mind that having insurance doesn't allow someone to fly like a maniac. Most policies will have reasonability clauses concerning the location of flight ("Who can stop me from flying at end of the runway of a major airport?") and the means that flight was being executed ("Check out how high my drone can go...I bet it is able to hit 20,000 feet easily!").

It is up to you but it may well be worthwhile to purchase some good liability insurance or check if you may have some coverage under your homeowner's policy if you fly as a hobbyist. Most homeowner's policies do not include this coverage but I can visualize that as the Dronespace becomes more popular, it may become an additional incentive for some companies to include it.

Hobbyist coverage of $1 million or more is actually very cheap. While this is an example and not an endorsement of any program, the Academy of Model Aeronautics (www.modelaircraft.org) provides $2.5 million personal liability insurance for non-commercial use as part of its annual membership (currently $75/year). The benefit to an organization like this is that it also advocates for your hobby.

Commercial liability coverage is available through many aircraft insurance programs. It tends to be much more expensive than hobbyist. At the time of writing this, it is about $500/year per every $500,000 of coverage. Of course, if you are using a drone commercially it is well worth the coverage as its cost is part of the cost of doing business. It will also help you remain competitive with others because a most venues you might be hired to photograph or video will require you to maintain liability insurance.

Safe and Legal Flight

Why Is It Important to Fly Safely?

If you have read through the previous sections, you have already picked up the gist of all that can be stated under this heading. It is not the author's intention to overstate this fact. It is important that all of the denizens of the Dronespace understand there is a lot of responsibility riding on flying safely.

One of the problems is that our nice shiny plastic and metal drones look so much like harmless toys that we don't grasp that they are potentially dangerous. Adding momentum from lateral and vertical movement makes it possible for them to gravely injure or even kill people. Flying a drone in a flight path could bring down an aircraft and cause tremendous carnage to the passengers and to people on the ground. Operating a drone in a location where rescue and firefighting operations are occurring could ground the very aircraft that are needed. Unplanned flight of a drone over highways and in cities could result in a crash that could hurt pedestrians and even may trigger a chain-effect vehicular accident. Improperly maintained drones can pose fire and crash hazards.

Hopefully, all of us will put aside the haphazard, front-page news-making early years of drone operation and seek to make flight safety our goal. The safer the Dronespace becomes, insurance rates will drop commensurately and regulatory organizations will not need to continuously slap on restriction after restriction. And it goes without saying, who wants their drone activities to lead to jail time, harsh fines, or guilt for killing one or more people?

Registering your Drone

Since the end of December, 2015 in the US, the FAA requires all drones to be registered as long as they are equal to or above 0.55 pounds on takeoff. However, early in 2017, a federal court ruled that the requirement to register non-commercial drones violated the FAA Modernization and Reform Act of 2012 and thus could not be enforced. That referenced law was scheduled to expire in September 2017, so the FAA may legally reassert the rule to register all drones, commercial and recreational.

They have streamlined the process through a single website found at https://registermyuas.faa.gov/ which is pictured below. The drone registration cost is $5 as of the publication of this book.

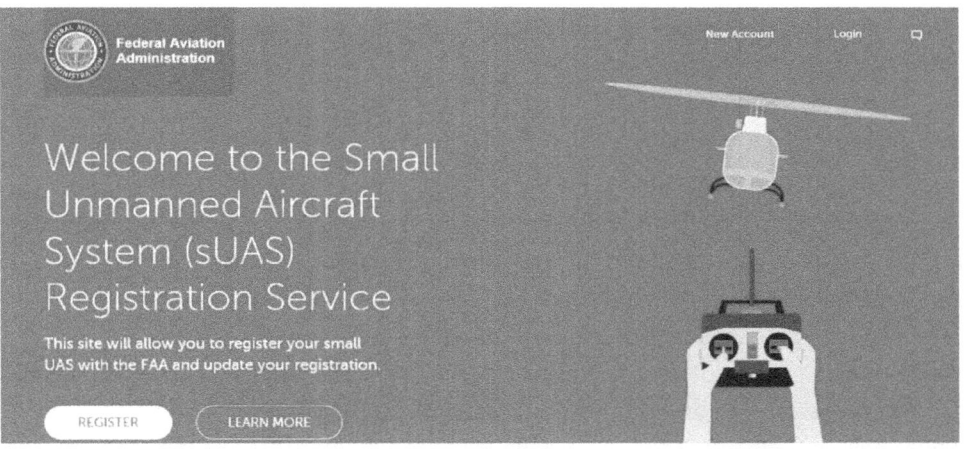

Do I need to register my Unmanned Aircraft?

You need to register your aircraft if it weighs between 0.55 lbs. (250 grams) and up to 55 lbs. (25 kg)

Figure 2 Screenshot of FAA sUAS Registration Website

Failure to register the (commercial) drone can result in penalties from the FAA up to $27,500 and criminal penalties of up to three years of imprisonment and/or fines of up to $250,000. It is quite clear that the FAA is not playing around and the first few violators that are caught will probably be made into examples.

Visitors to the US travelling with their drones are also required to register their drones. This registration varies upon citizenship. If a visitor is a US citizen, they can register using the above site. Otherwise, the FAA's process for any foreign national who might "want to operate your UAS exclusively as model aircraft you must complete the steps in the web-based registration process and obtain a 'recognition of ownership.' This recognition of ownership is required by the Department of Transportation to operate a model aircraft in the United States." Commercial drones owned by foreign nationals intending to be used commercially in the US need to be registered in their country of origin and then that registration must be transferred for operation in the US through the Department of Transportation. Consult the FAA's website for the latest details in either case.

Once the drone has been registered, the FAA requires that the registration number be legibly written on the drone's airframe using permanent marker or some other system that will not come off during normal flight and handling of the sUAS.

Respecting No-Fly Zones

The rules for flight are slightly different between hobbyist use of the Dronespace and commercial users. There is a clear overlap of these rules for both of these uses, especially in terms of No-Fly zones. For example, Restricted Airspace is restricted for all drones regardless of intended use. Commercial and non-commercials drone pilots are not allowed to fly over the Mall area in Washington, DC. Commercial and non-commercial drone pilots are also not allowed to fly on airport tarmacs, although for some special reason a commercial drone pilot may be granted a limited waiver.

The same rules of No-Fly apply to commercial and hobbyist flight over any property owned by the National Park Service (NPS).   Until further notice, there is a ban on drones in place on National Parks and National Monuments that is lifted only in the case of NPS-sponsored drone use for animal population studies and such internally-driven endeavors. Other National Park Service assets such as National Seashores are also protected. On the positive side, there is supposed to be a process in place to seek written permission to fly for justifiable reasons from the superintendent for NPS properties.

Note that U.S. Forests can be used by non-commercial recreational drone operators per their published policy (https://www.fs.fed.us/science-technology/fire/unmanned-aircraft-systems) and can be used by commercial operators if the seek and receive a permit.

Be aware that sometimes protected NPS assets are not readily recognizable as such. In 2017, while visiting Alaska on a cruise, we stopped in Skagway to discover most of the town itself is part of a National Park Service's Klondike Gold Rush National Historic Park. After speaking with a very friendly security guard who directed us to a point that was outside of the (unmarked) No-Fly zone, we were able to hike to that area and pop our drone up to grab some video and shots.

Bear in mind that flight over US territorial waters is also under the control of the FAA and other agencies. For example, it is illegal to fly a drone in National Marine

Preserves because there is an established flight minimum altitude over wildlife of 1,500 feet AGL (Above Ground Level). Since drones are limited to a maximum altitude of 400 ft AGL, they can never be used legally commercially or not in these areas without special permission. Even in the open ocean, it is illegal to overfly marine animals or nesting birds. Whales have a 1,500 ft minimum no-fly cap over them that extends 300 feet in radius around them. Again, drones cannot legally fly at this altitude. Even manned aircraft (planes and helicopters) are limited to a single pass at this altitude over a whale.

There are No-Fly Zones that the federal government, state governments, and local municipalities have established. The FAA has released a smartphone app called B4UFly for iPhone and Android devices which outline many of these zones. It is incumbent upon the pilot to confirm the fly and no-fly zones based upon the rules for hobbyist use or the rules for commercial Pilots-in-Charge depending upon your desired usage. It is advised that the relevant questions and answers at https://www.faa.gov/uas/faqs/ be scrutinized by a non-commercial drone operator.

The rest of this chapter is mostly targeted to the commercial drone operator although there are some good practices even a non-commercial hobbyist might want to incorporate into their operating procedures. This is especially true if you have any intent of becoming a licensed Pilot-in-Charge later.

Roles on a Drone Team

In Part 107, the FAA outlines several roles on a team that operates a sUAS commercially. It is crucial for the reader to grasp that one person may hold multiple roles or multiple people may hold a single role on a team. There are also roles that the FAA does not mention but which we need to be aware of.

The critical role is that of Remote Pilot-in-Charge (PIC). This is an FAA-required role for a drone to be used in any commercial venture. All responsibility for the preparation, legality, and success of the flight rests on the PIC. A PIC must be licensed as such by the FAA. Merely holding a regular pilot's license does not qualify. A licensed pilot must take a test and be specifically licensed as a sUAS PIC. The PIC plans out the operations for the drone, checks for airspace restrictions, creates NOTAM (NOtices To AirMen) if needed, secures ATC permissions if needed, inspects the drone and its loading, briefs the crew on the mission, prepares emergency procedures, and monitors the operation while making changes as needed to accommodate changing conditions or emergencies.

The person who controls the drone's flight using the Control Station (CS) is known as the Person Manipulating the Controls (called Controller in this book). This may, or may not, be the same person as the PIC. If it is not the PIC, the PIC must be in position to take the control station from the Controller in case he or she needs to take control during an emergency. The Controller must be constantly either in direct line-of-sight of the drone or must be in direct contact with someone who is. The FAA is clear on the fact that this line-of-sight must be unaided by optical or electronic means (other than corrective lenses). Using FPV (First-Person View) goggles is not line-of-sight.

Any remote person who is in charge of maintaining line-of-sight with the drone and hazards within its airspace while it is out of the Controller's sight is called a Visual Observer (VO). The FAA is strict that the VO and the Person Manipulating the Controls must have a constant contact while the VO is in control of monitoring the aircraft and monitoring for hazards and other aircraft. The VO is also restricted to unaided vision.

While the FAA recognizes these roles, there are others which may comprise a functional drone team. The first of these is a Mechanic. Most small drones are maintained by their owners but larger drones may need a dedicated person to properly maintain and prepare it for flight. Another role which may also not be obvious is a Loadmaster. Again, small drones less than 15 pounds may be loaded by their owners or operators but the larger drones may need to be loaded and properly balanced by a professional.

A role which is not named by the FAA but which is as critical to a commercial drone team is that of Payload Operator. In most cases, a commercial drone is not flown simply to be in the air. That drone's operation invariably is driven by the payload. The most common Payload Operator is a Camera Operator who controls the framing, settings, focus, and capture of photos and/or video. Many serious drones now have the capability for a Camera Operator's station as an option. Other drones have specialized sensor equipment as their payload such as Infrared cameras and the Payload Operator for such equipment is generally trained in using and interpreting the sensed data.

One of the tasks of the PIC is to incorporate each of these roles into a mission, to separate roles as needed, to add additional VOs if required by the operational envelope of the mission, and to brief everyone so they are clear on the expectations of their roles. Depending upon the complexity of a mission, the size of the sUAS involved, and its loading requirements, the team can be anything from one person to a dozen or more.

Using Aviation Planning Tools

There are a number of planning tools that exist to assist regular pilots to plan their flights and a number of tools that have arisen to satisfy the new drone market. Therefore, there is a mix of traditional paper-based tools and smartphone tools at the fingertips of any aspiring drone pilot.

If you are merely flying as a hobbyist, your role is to communicate to any airports or heliports within a five mile radius of your flight zone and to ensure that you are not trespassing into any No-Fly Zones. The FAA has released a smartphone application called B4UFLY for both iPhone and Android platforms. This will quickly show you the geo-fenced areas surrounding your chosen flight location. Here is a pair of screenshots from the Android-hosted application in my hometown.

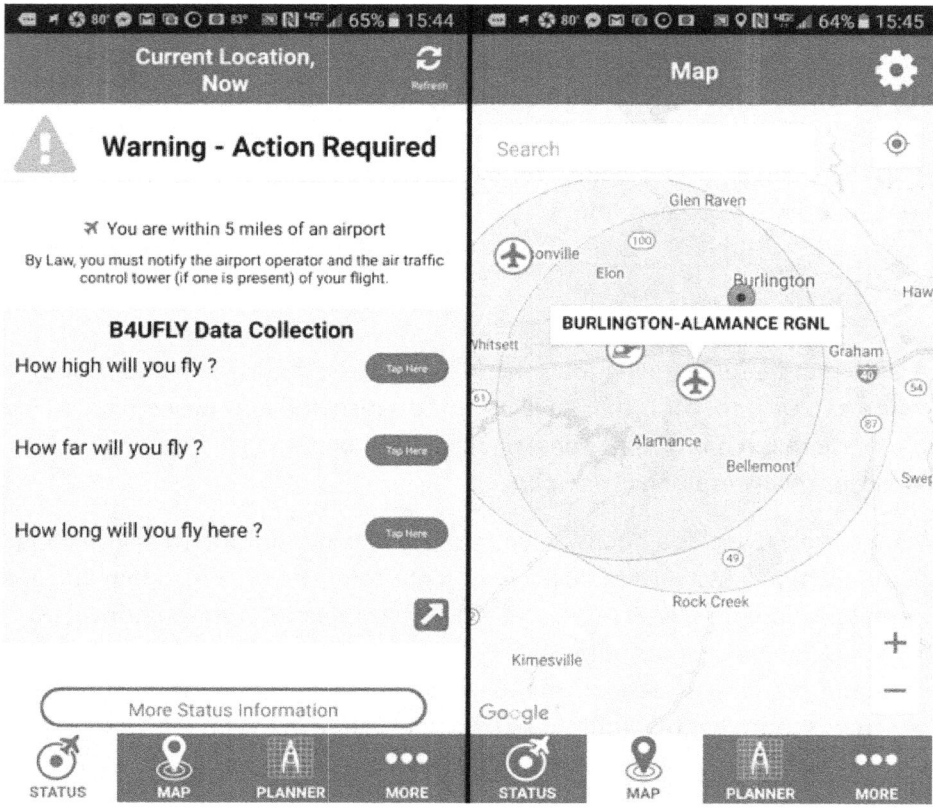

Figure 3 FAA's B4UFly smartphone app

The easy-to-read map indicates that my location is in the 5-mile range of two airports. Clicking on the name of the airport brings up information about that airport. The rule for non-commercial flying (as you can see on the left-hand screen) is that you have to notify the airport manager. Here is where I think that they dropped the ball. The information screen for the airport does not provide you with the phone number to make the contact so you have to scramble to find it elsewhere.

There is a publication called the Chart Supplement that contains all relevant information for all airports that are open to the public. There are also a number of smartphone apps that tap into the information in the Chart Supplement data and pop it up onscreen. I use one on my Android called FlightIntel. Using something similar to this alongside the B4UFly app is helpful to find information such as the manager's number as you can see in the lookup for Burlington-Alamance Regional.

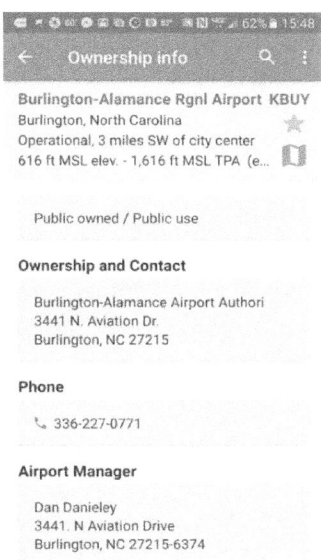

Figure 4 Airport Ownership information from FlightIntel

A call to an airport manager is nothing to cause concern. You just need to introduce yourself and let them know that you are calling to inform them that you will be flying a drone at a maximum of however many feet at a point located however many miles in whatever direction from the airport. You will want to let them know the timeframe of the flight.

The FAA indicates that the manager cannot prohibit flights arbitrarily. Their specific language is that "an airport operator can object to the proposed use of a model aircraft within five miles of an airport if the proposed activity would endanger the safety of the airspace. However, the airport operator cannot prohibit or prevent the model aircraft operator from operating within five miles of the airport. Unsafe flying in spite of the objection of an airport operator may be evidence that the operator was endangering the safety of the National Airspace System. Additionally, the UAS operator must comply with an applicable airspace requirements." (www.faa.gov/uas/faqs) If the manager objects, it is imperative to understand the objection in terms of aircraft safety. Perhaps a compromise solution can be worked out by asking if an alternate site is appropriate for flight instead (assuming that site is still within the 5 mile limit)

Commercial drone operators have a great deal more homework to do. Many PICs would prefer to map out their airspace usage using a standard FAA Sectional map to ensure that they are going to operate within Class G airspace, check on weather conditions, check TFRs (Temporary Flight Restrictions, which should also be done by hobbyists using B4UFly) and NOTAMs. Sectional charts are published and available through the FAA in paper form and also in PDF or TIFF forms (see www.faa.gov/air_traffic/flight_info/aeronav/digital_products/vfr/ for more information). Sectional charts permit a complete understanding of the makeup of the airspace, including restrictions and obstacles such as towers that might be intruding into that airspace, and weather/control tower/UNICOM radio frequencies for airports in the area. There are also a number of applications available for smartphones which provide instant access to sectional chart information. My preferred tool for this is Avare and a sample screen of the Burlington-Alamance Regional airspace is pictured here.

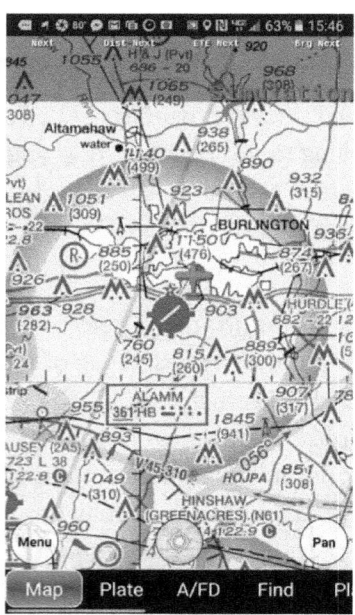

Figure 5 Avare screenshot showing KBUY (Burlington NC)

Avare, like many of its counterparts, also provides an interface to search for an airport and see the Chart Supplement information such as runway information, management, and phone numbers.

For a PIC, getting updated airport-centered weather is important. This can be done using an aviation radio (see below) is within range of an airport, or by using online access tools. Web-based tools such as Weather Underground permit you to view the METARs (Meteorological reports) from an airport. There are also lots of airport weather apps for smartphones such as Avia Weather which I use personally. All of these sources permit viewing the raw METARs as well as human-interpreted versions of the same.

Airmap is an app available for use both on Apple and Android platforms. It brings together the concepts of airspace and flight planning into a single application. This is critical from their perspective since Airmap is a leader in the LAANC space (described later in this book) so precise planning may be needed for requesting access to one or more squares within a LAANC grid. Airmap permits the user to view airspace as a recreational pilot, an Part 107 pilot, or a 333 Exempt pilot.

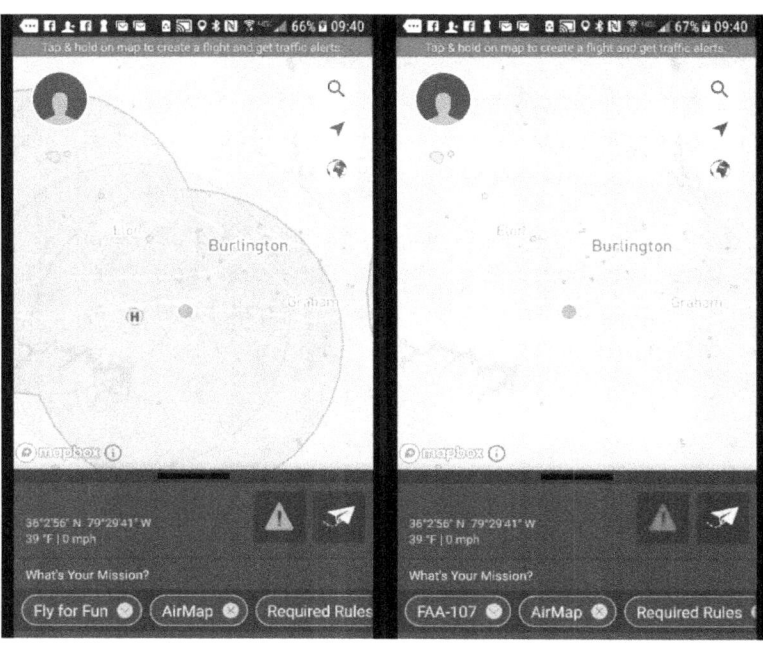

Figure 6 Airmap viewed as Recreational and Part 107 pilots

Once a flight location has been selected, Airmap then permits the pilot to lay out a flight plan and once complete, it files a NOTAM and, in the future, any LAANC requests that need to occur.

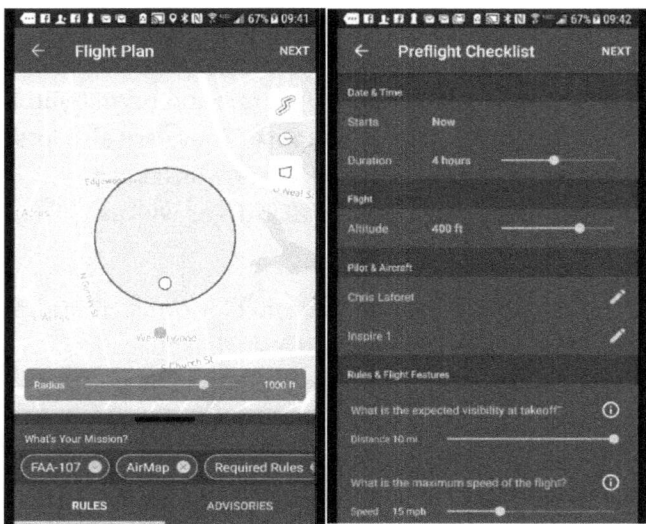

Figure 7 Airmap flight plan mode

Finally, there are apps that support pulling NOTAMs and TFRs. Some of the apps already mentioned in this section do this as part of their offerings. There are also other smartphone apps such as NotamDroid that may be handy.

The website www.1800wxbrief.com (and its dial-in at 1-800-WX-BRIEF) is an indispensable source for all the above information in a one-stop shop. Some PICs prefer to do all of their planning in that one location which now provides a specialized planning section just for UAS pilots.

When Do You Need an Aviation Radio?

In most commercial sUAS operations in class-G airspace use of a dedicated aviation radio is not required. However, in operations within the vicinity of an airport or specialized operations approved by ATC (Air Traffic Control) such as within Surface Class-E, it may be important enough to maintain the possibility of a two-way communication. Entry level two-way handheld aviation radios are not extremely expensive.

Aviation radios are also helpful to monitor CTAF frequencies of nearby non-towered airports during a mission. This permits a PIC to determine if the traffic patterns change unexpectedly and to be able to make announcements on CTAF to advise pilots who may be straying near to the location of the mission to be aware of the drone activities.

Thus the decision to invest in a radio is definitely personal. It depends on where you will do most of your flying. If your activities are centered on areas that tend to be dense with small CTAF-traffic airfields, you may be safer making the decision to make the purchase.

Filing for Waivers from the FAA

Whenever you have a scenario that does not fit into the standard rules in part 107, the FAA permits for filing a waiver exception. It is advisable to file these at least 90 days in advance of the date the exception waiver is needed. The waivers are filed at https://www.faa.gov/uas/request_waiver/. Note that you cannot request waivers for every aspect of part 107. The form at the site provides the list of waivable regulations.

When you go through the process, you must describe the conditions of the waiver and outline all the interventions and processes you will take to mitigate the risks that the exception is going to cover. Be aware that your request may affect more than one regulation so make certain you cover all of them.

Think "ironclad" concerning how you will deal with each and every risk. Don't jump in and hope for the best. The FAA will scrutinize your processes and will deny your request if anything does not sit right with them. Bear in mind also that whatever you are attempting to request a waiver for must be completely accounted for in your supporting documentation in terms of protection of people, the lighting of the drone, the loading of the drone, and other modifications that may be necessary to be done to the drone and operating policies to make the operation safe.

Coming Soon to an Airport Near You: LAANC Authorizations

In 2017, the FAA along with several drone mapping providers (led by AirMap) have released a new system entitled LAANC. This acronym refers to the Low-Altitude Authorization and Notification Capability system which will open up an automated process for requesting access to controlled airspace.

The goal of LAANC is to be able to dynamically request access immediately to small sections of controlled airspace (e.g. Class B, C, D, surface-E) instead of having to go through the tedious waiver process. The goal is to simplify commercial operators' needs to fly their low-altitude missions within the vicinity of major airports. The system will roll out piecemeal at some test airports in 2017 and then be extended in 2018. Its first application is in Class E airspace and will extend into the other classes as the program expands.

LAANC is built upon the FAA's UAS Facility maps process which is documented at https://www.faa.gov/uas/request_waiver/uas_facility_maps/. Users of AirMap (documented elsewhere in this book) and other LAANC software will eventually be able to request ad-hoc access to a grid square at the maximum altitude permitted from that software. This is a really exciting improvement which will make commercial drone operations more effective and efficient.

Flight and Maintenance Logging

Commercial sUAS operators should keep logs on flights and significant maintenance items. Even though hobbyist drone use does not require logging, it is also a good idea to keep a logbook. Many flight applications such as DJI Go do track flights and log them. This approach may be good enough for most. Others would argue that there is no guarantee that the manufacturer-maintained logs are guaranteed to be around for the long haul. This is a powerful argument especially when we consider why logs are important.

Consider what will happen if there is an incident involving you and your drone? If you are a commercial drone operator, the FAA may require you to produce flight and maintenance records. The logs maintained by a manufacturer's program may track the flights but not the maintenance you have done (at least, as of the time of writing this book). It is not a bad idea to maintain a separate copy of your logs to cover yourself.

There are apps for iPhone and Android devices that allow you to track flights and/or maintenance. There are also paper-based logbooks which some prefer to use. Some (me included) track flight and maintenance in Excel spreadsheets. The key is to try several options and see which works best for you. Obviously, the solution that is easiest to use will tend to be consistently used.

Maintenance logs should track routine and other operations that are performed on your drone. Maintenance specifications may be published by your drone's manufacturer and/or may be supplemented by reasonable procedures. These would include off-cycle procedures such as software updates to the drone and controller software and replacement of propellers and batteries. In my log, I track every software update and normal maintenance. Tracking the time and money spent on these items is helpful statistical data. Additionally, on my flight logs, I even track which batteries (I have tagged each with a different letter) are used so I can ensure they are rotated through evenly.

Learning to Fly Rotorcraft

Understanding the Lingo

As stated earlier, this book mostly centers on multi-rotor drones. There are drones that are modelled after conventional fixed-wing aircraft also and these are discussed in their own section. So, much of the lingo and techniques for flight in this section concern rotorcraft.

Drones, like any aircraft, fly in a three-dimensional space. Movements in 3-D space are controlled by a controller with directional controls which trigger each specific movement. Aircraft have specific movements along each axis. These axes are aligned at 90 degrees to each other (see figure 6) in the longitudinal, lateral, and vertical orientations. The movements are, in order of the aforementioned axes, roll, pitch, and yaw.

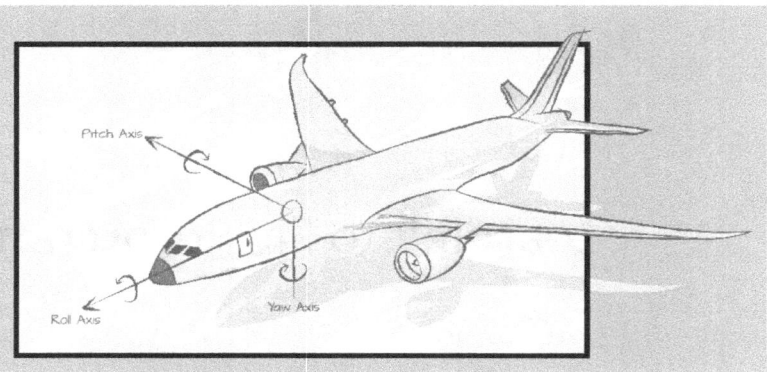

Figure 8 Movement around Axes of Flight

Unless one is flying in sport mode on a high-end rotorcraft drone, the only axis that is active is yaw. Most of the larger drones tend to keep locked in level flight mode, which is to say that the craft reduces the effect of roll and pitch to a minimum. Controllers normally provide inputs to control these two axes but the net effect for the pilot is not a pitch or roll but a movement forward/backward or laterally to the left/right. If the craft has a camera mounted to a gimbal, that gimbal ensures that movement on these axes is completely nulled out so that any footage captured by the camera is rock steady. Since the camera generally feeds POV imagery (Point-of-View video transmitted from the front of the drone to a screen on the controller) to the pilot, that imagery is also rock steady, at least on the roll and pitch axes.

Mastering the Controls

A multi-rotor drone is controlled generally by using a controller. There are several different forms these controllers can take. The simplest option is a drone that is controlled simply by using a WiFi signal from a phone or tablet device. These types of drones are restricted on how far they can travel from the controller's position before they lose signal. Thus, this form of controller is generally only found on small toy drones. There are some exceptions such as the DJI Mavic which can use WiFi up to 50-80 meters or use a dedicated controller for larger distances.

Most higher-end drones feature a dedicated controller with a much better range than plain WiFi. Some of these controllers have built-in screens to provide POV video and others provide a dedicated port to connect to a smart app on a phone or tablet. Either way, unless set up otherwise, these controllers feature a left control that commands lift and yaw and a right control which commands forward/backward and left/right horizontal flight (see figure).

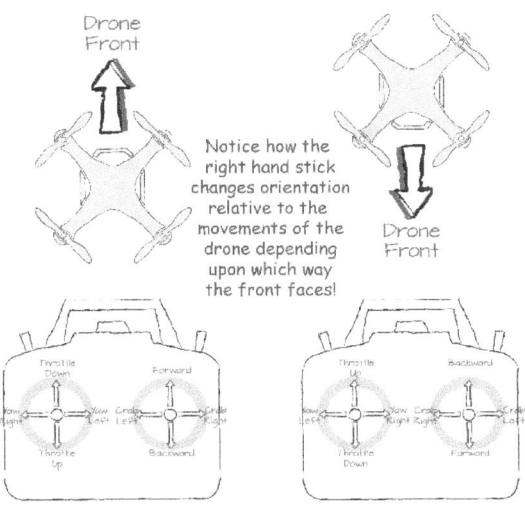

Figure 9 Multi-rotor drone controller (Type 2) layout

It is important to gain an understanding of how these controls work. Picking up a controller the first time can be somewhat intimidating. This is even more so if the drone either does not have GPS attitude positioning support, or if it cannot get a GPS signal, or is being run in a non-GPS mode. The nice thing about GPS support is that when a drone is in flight, releasing the controls merely lets the drone hover in-place without any drama. Without GPS control, the operator will have to constantly control the position of the aircraft by juggling these controls.

So, it seems logical to just buy a GPS-enabled drone and pop it into the air without having to really know how the controls work. What can possibly go wrong? Lots! While it may make sense for a beginner to fly really low and slow in a large controlled environment (no trees or structures around) until they get a handle on the controller's functions, it certainly is not wise for a beginner to try to fly some sort of mission without a good feel for what the controls do. What happens if the GPS lock fails? What happens if the power hits a critical level and the controller has been set to hand control off to the operator? What happens if you have to do a controlled emergency landing? As with all things that fly, there are an infinite amount of scary scenarios that can occur out of the blue.

Don't let this last statement scare you so much that you decide to not fly a drone at all. Flight is a combination of science and art in which a pilot anticipates as many risks as possible and attempts to mitigate them one by one. The science part involves understanding everything about flight and the environment in which flight occurs as possible. The art side involves learning how to deal with the controls so well that actions and reactions are almost instinctive in nature. As part of the art of flight, a drone operator should practice flying with and without GPS support and be able to juggle the controller functions effectively. Some of this can be developed early on and the rest will come with diligent practice over years of operation.

As you can see from the diagram above, the altitude of the drone is controlled by using the left stick and moving it forward to make the drone climb faster or backward to make it lose altitude faster. Generally the center position is a spring-loaded détente where nothing changes in the altitude of the drone. Pulling the stick backwards is the way you can bring a drone in to land. It is important that you check your user's manual to know how to turn on and off the propellers on the drone. Some use a combination of movements on the control sticks, others just sense the left stick being held down for some period of time to turn off the motors.

The same left control stick also controls the yaw of the drone. Moving it to the left will make the front of the drone pivot leftward and continue rotating it counterclockwise until the stick is returned to the détente in the center. Moving it to the right moves the drone around clockwise instead. The behavior of these controls will never change regardless of which way the drone points. Be aware, however, that when you rotate the drone's front to point in any direction other than facing the same direction that you are facing, the behavior of the controls on the right stick will appear to change.

When the front of the drone is facing the same direction as you, moving the right control stick forwards will make the drone move away from you. Pulling it back will bring it back towards you. Moving the stick to the left will make the drone crab left but keep the same distance and heading from you. Moving it to the right will make it crab right. So far, so good. The best thing for most beginners who are flying a drone is to always keep the front of the drone heading the same way they are facing. In this orientation, the controls will always make sense.

Things get dicey as soon as an operator decides to yaw their drone left or right. Imagine yawing the drone 90 degrees to the left. At this point, pushing forward on the right stick will not move the drone further from the operator but will make it crab to the left! This will make sense if you think that you are moving forwards but the front is pointing left now. In order to move the drone further away from the operator would need the right stick pressed towards the right! This is one of the reasons that it was stressed earlier to really get to understand the controls. What if your drone had a tree branch just off to the left and it was yawed to the left. If you wanted to avoid the branch by moving it further away from you and you pressed the right stick forward, you would smack your expensive drone into a tree branch!

Some drones offer "headless"/IOC/Intelligent Orientation Control modes that permit the right hand stick to always work correctly regardless of where the drone is to the operator. Your user's manual will show you how to set this up and it is not a bad idea if you are a new drone operator. These modes depend upon GPS and knowledge of where the controller is or the take-off point was to convert a right control stick input into a movement in the intended direction. For example, in the case above with the tree branch, pushing the control forward would crab the drone to the right away from you and avoid crashing into it.

A rule to embed in your mind is that if you ever lose a sense of direction, just stop the drone's movements, use the left hand yaw control to rotate it to point away from you, then you can control it normally with the right hand control stick. Most accidents happen when an operator is confused and reacting to a situation. By stopping and taking control of the situation you will get the time to think and to process things normally.

A Training Drone

In light of the importance of learning how to use a drone's controls effectively, many drone advocates (and me included) recommend purchasing a cheap toy drone. For $40-$60, there are many good little drones that will take a licking and keep on flying. These little drones (mine is a Hubsan X4) allow you to practice indoors, can be easily and cheaply repaired, and will provide invaluable flight skills.

Many of them do not have stabilization and position-holding systems which make them very challenging to keep flying level and require your total concentration. In fact, one of the best aspects of flying them is that you learn to be gentle yet definitive with control stick movements. This is especially true with the left-hand power stick. Since these drones tend to be a bit underpowered, making any other move (forward/backward, sideways, or yaw) tends to need a gentle addition of power to keep the drone's altitude. You will find that adding too much power on the stick can shoot the drone up towards the ceiling and your native response is to chop the power. This has the immediate adverse reaction of sending the drone plummeting to the floor. Over time, you will learn how to moderate adding power for a more gentle climb and if the drone starts to climb too fast, to ease back carefully on the throttle instead of chopping it.

Starting with one of these small, inexpensive toys and flying it around for a few weeks before firing up a big drone is well worth it. They will get you thinking more critically about the decisions you make while you are in the air, the consequences of bad decisions (In my case, once I was flying through the kitchen, lost control, and plummeted into a sink filled with dish-water! I experienced a

"Doc" from <u>Back to the Future</u> moment…), and that flight is never simple even if technology makes things easy. When you fire up the larger drone and first get it into the air in all of its GPS-stabilized glory, it will suddenly seem so much less intimidating than the little toy!

There are a number of great exercises you can work on with the toy drone and eventually work up to doing them with the larger drone in ATTI (Attitude) mode (which turns off GPS controls) as you gain confidence. One of these exercises is to mark out 4 circles just slightly larger than your drone and placed as four corners of a box. The objective is to lift off out of one of the circles, fly to another, land smoothly, then repeat by flying to the other circles. The circles can be close together initially and can be moved farther away from each other as you gain more skill. Later on, rotate the drone to a different angle (45 degrees, then 90 degrees, then 180 degrees) and fly accurately with the right stick from landing-pad to landing-pad.

Another exercise is to create obstacle courses and fly through them without mishaps. Over time, you can try flying the courses with the drone pointing in a direction other than away from you. With larger drones, a course can be set up with photographic lighting stands or something similar. Since the stakes are higher and damage can be more expensive with these drones, take it slow and easy so you can process what is happening until you fully understand what is going on. Run exercises with GPS enabled until you feel confident enough to try ATTI mode, and always be ready to click on GPS if there is the slightest problem.

Making Flight Easy with GPS, Sonar, and Pattern-Matching Sensors

In the last section, we touched on ATTI mode. ATTI stands for Attitude and in the case of DJI drones, it means that all sensors are disabled except for the atmospheric pressure sensor. This permits the drone to keep more-or-less at the same altitude when you hover but there is no other support for locking its position.

I don't know of any higher-end drones that do not include GPS tracking. Some, such as the DJI Inspire, actually have GPS in both the drone itself and in the controller so that the software can accurately determine how far away and in which direction the drone is relative to the operator. GPS support ensures that the drone remains locked to a specific location and is used also in a return-to-home scenario so that the drone can come back and land where it took off from in case of an emergent situation such as low battery power. GPS support also is used by the software to lay out a track to follow or a position to orbit and film. We will get into these functions later on when we delve into filming techniques.

Some drones also use sonar sensors to sense their altitude from the ground or to prevent crashing into an object in its path. Pattern-matching optical sensors also are used to avoid collisions and can also be used to precisely hover in locations where there is not a GPS signal (such as within an indoor venue). These sensors look for patterns on the ground and attempt to maintain orientation relative to those patterns.

All of these are handy features and, depending upon your planned use for a drone, may be worth the extra cost.

A Short Discussion about Cameras

When you are considering cameras, there are a wide range of options available. Some drones have built-in cameras which cannot be replaced. Others offer several options at different price-points and physical capabilities. Some high-end professional drones are sold without any camera options because they are payload-agnostic platforms that are adaptable to cinema cameras such as Blackmagic, Red, Arri, etc. In this case, the weight and power requirements of the cinema camera package and the performance envelope of the drone itself will dictate the decision of which drone to purchase.

If quality images are your goal and you are already involved in filming and photography, beware of lofty camera claims. Just as there is an ongoing onslaught of numbers in the regular camera market, the same is true of the drone camera market. Cameras are presented as capable of high-megapixel stills and large-K video data. These numbers are as meaningful or meaningless as they are in the rest of the camera world. The quality of the sensor and the precision of the lens is more important than some number. It is advised that you spend time reading and watching reviews of your desired camera before committing to it.

Some drones use a fish-eye ultrawide angle camera such as a Go-Pro and there are others which use normalized lens cameras. Some options offer interchangeable lens cameras, and some options offer higher-end photography and video RAW and/or Log captures.

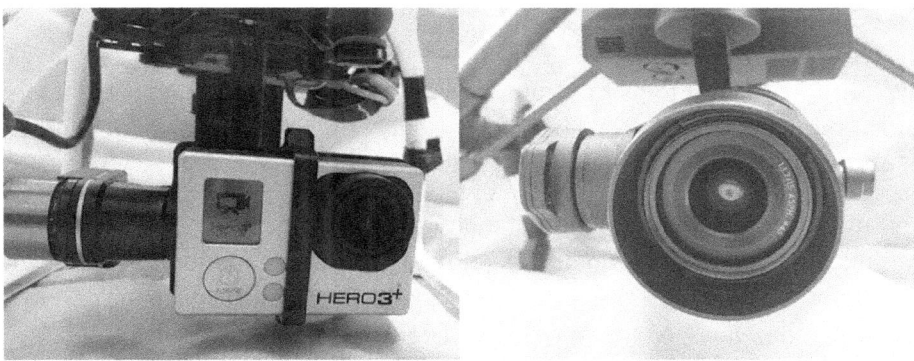

Figure 10 Two DJI camera options. A Phantom 2 with a Go-Pro Hero 3+ and an Inspire 1 with a Zenmuse X5-RAW

In the attached figure, here are two of the author's drones. The older Phantom 2 was purchased for general use and for some filming with what was at the time the state-of-the-art GoPro Hero 3+ Black. The newer Inspire 1 was purchased very specifically for commercial photography and videography with the highest option of three camera offerings. The Zenmuse X5 is a four-thirds sensor camera with interchangeable lenses that shoots RAW 4K video and 16 Megapixel images and offers a 12+ stop dynamic range. Both cameras are very useable for a wide range of shooting but quite obviously the X5 provides capabilities better suited for high-end visual work. There are pros and cons to each of them starting with initial cost, replacement cost risk, image quality, processing options, and so on.

Before getting off the subject of cameras, it is just as important to consider how your images and video will be processed. The formats shot by the cameras vary in their need for resources (e.g RAW video needs a lot of disk space and computing power) and editing capabilities. A simple JPEG photo can be uploaded directly from the camera to a website whereas a DNG image may require processing with a package like Photoshop before it can be used. The same is true of video capture. Some cameras capture video in one of the common H.264 codecs and are easy to edit with simple tools and upload to YouTube, Facebook, or Vimeo.

Others use other formats that are meant to be professionally edited using packages such as Final Cut, Premiere Pro, Vegas, or Davinci Resolve. The decision for what camera you choose will hinge on these post-capture details.

Choosing a Drone for Yourself

So, now it is time for you to pick a drone for your use. What should you look for? How much should you budget on spending? Is any one brand better than another one? These are all good questions, for sure.

First and foremost, you need to understand what you specifically need this drone to do. Is it just something for fun or is it going to become a commercial drone? A fun, hobby drone could span the range of pricing from relatively inexpensive to moderately expensive. A commercial drone needs to be robust, generally has some strict requirements that it needs to meet, and will generally span the moderately expensive to very expensive pricing range.

Many drones are purchased so that they can be used to capture photographs or video. There are drones, especially commercial drones that are merely flying camera platforms. The core object on these drones is the camera itself and everything else plays second-fiddle to that camera. Other drones are purchased for flight purposes and the cameras are more of a second-thought. Ask yourself where does a camera fit in your drone equation? If the objective for owning a drone involves high-quality photography and video, then the camera options will dictate in some large degree what drone options you can consider.

While still on the subject of cameras, these are normally mounted on gimbal mechanisms so that they remain rock-steady as the drone moves around in 3-D space. Most gimbals are 3-D but some cheaper options only work in two dimensions (roll and pitch). If videography is one of your goals for a drone, do pay careful attention to the gimbal options available.

Another neat feature that may be a deal-maker or deal-breaker in deciding on a specific drone platform may be if it supports dual controllers or not. A dual-controller setup can cleanly split up the drone's operations into its key components. One controller is used by the flight operator in controlling the flight's mission while the second controller can be used to control the orientation, parameters, and actions of the camera. This also permits a real photographer or videographer to be in complete control of the creative side of the drone's use while an experienced pilot/operator can concentrate unhindered on flying the drone.

On the other hand, you may be looking for a racing drone, something that you can use to tear open a rift in the space-time continuum as you set new speed records. In this case, you may be interested in a camera on the drone solely for its FPV feed but nothing else. What bubbles to the top of your list is raw, unadulterated linear speed.

There are a plethora of drone manufacturers out there now, many of whom use proven and standardized chipsets in their drones. One of the largest is DJI and, in full disclosure, I own two DJI drones. My wife owns a DJI also. Some others out there at this time include Yuneec, GoPro, Parrot, Blade, Syma, and Hubsan. Different manufacturers will be found competing at different price points. Drones are sold directly by manufacturers, in online stores, or even in some big box stores such as Best Buy.

The key is to hone in on what you need and then discover which drones get you there. As with all things nowadays, it is a good idea to scour online reviews before committing to the purchase. Many cities have meetup groups for drone enthusiasts and it is worthwhile going to a flying event and ask some pointed questions (You can look for drone groups in your area on meetup.com).

Two Critical Drone Innovations in 2017

The Superbowl 51 audience was introduced to drones in an innovative halftime show featuring Lady Gaga. Ascending Technologies, a German company that was purchased by Intel the previous year, demonstrated to the public the power of swarming drones. In the display which used some 300 drones with lights, the swarm executed patterns and ended with 3D illuminated image of the US Flag over the stadium.

The drone spot was actually filmed a few nights before the actual Superbowl was played since the technology is new and it operated under waiver from the FAA which permitted multiple drones to be under the control of one pilot and specialized geofencing software. This software ensured that the drones' behavior could be choreographed and contained within a defined 3D space. It was a positive presentation of drones to the public. Needless to say, drones were the center of many watercooler discussions for the next few days.

Also during the first half of 2017, the drone design world changed in an instant. DJI released a new drone that built out on the basic design of the Mavic but which cemented a new direction in drone design. Its innovations will ripple into drone manufacturing into the future. The Spark is a small drone, fitting very comfortably in the palm of one's hand. It is small and it is relatively inexpensive but it is anything but a toy. Its small battery can keep it aloft for a solid 15 minutes!

It can be flown in one of three ways, either through use of an add-on traditional controller, or using an app which provides on-screen controls, but also it offers a new third way to do basic flight operations. Gesture control/FaceAware permits someone to start, take off, move around, take photos, and land the craft back in the palm of their hand without once touching a controller!

The Spark also is an extremely stable craft for its size. Basically it offers three modes of operation using DJI's Go 4 app and/or external controller: The normal mode makes its movements slower and more deliberate – stabilized and controlled. Turning on Sport mode unleashes a beast that will rip along at about 30 mph. Then there is Tripod mode, which is also available in the Mavic and Phantom 4s. This enables a very slow mode that permits buttery-smooth movements optimized for filming.

Drone design is likely to build upon these two new technologies which will make them smarter, smaller, and more accessible to the public. Swarming technology also opens the door for new commercial, first response, and (of course) military applications.

Caring For and Feeding Your New "Baby"

Once you have committed to a purchase and bring home a new drone, what do you need to do to keep it healthy? Most drones are well designed and have very few needs. Do check your user's manual for any maintenance milestones that need to be followed.

It is advisable to order an extra set of propellers for your drone and always carry them around. It is easy to nick a prop blade so it may need to be replaced before flying again. If you have one with you, then your flying time will not be cut short by this maintenance need. It is a good idea to clean the blades with a damp cloth and inspect them before each flying session. If in doubt, throw it out! Do not skimp on a blade that may be less than perfect. It is the only thing keeping your baby in the air so it is not worth risking making your drone into a meteorite just to save the money on replacing a blade!

Also, unless you have experience with it, it is not worth getting blades from aftermarket suppliers. There is a lot of cheap junk out there interspersed with good blades. Unless you know from experience or from the experience of someone else that a blade type is excellent, just play it safe. Again, your expensive drone could crater because you decided to save a few cents. Ouch!

It is also worth purchasing one or more extra batteries for your drone. Again, I urge you to purchase from the manufacturer and not from the aftermarket unless you know what you are doing. For one, lithium ion batteries are seriously dangerous so you really should get the ones that are backed by an established manufacturer and not some fly-by-night operation. It is a good idea to label your batteries (I use letters, O for the original that came with the unit, then A-Z for others) and track them in a log so that you are sure that you are rotating through them properly. Do not ever trust your batteries to perform properly all the time!

True story time: My only crash was due to a bad battery and if I had been properly logging the use of my batteries earlier, I would have anticipated problems with it. I had a battery for my DJI Phantom 2 that seemed to charge fine, however whenever put under the load of flying, it would last a couple minutes then signal a low-battery condition. I did a test flight one morning over some trees and before I knew it, the drone was returning home with a low-battery condition. I had not anticipated this problem and had a RTH altitude that was just below what I should have set it for the tree height in this location and the drone smacked into a branch and crashed hard. On the old Phantom 2, the RTH altitude had to be set using a computer and a USB cable so don't be too hard on me. Life is so much easier now it can be done immediately using the tablet attached to the controller...

While on the subject of batteries, do not leave batteries charged up. The best approach is to charge batteries before you need them and to charge only as many as you will need. The new DJI intelligent battery design, and I am sure others have followed suit, has logic to auto-discharge the battery after 10 days of not being used so this is less of an issue in this case. However, to discharge themselves, they do generate some heat so this leads to another pointer: Do not leave fully charged batteries in an enclosed space for days. Take them out and put them on a counter or a shelf so they can remain cool if they auto discharge.

One final word about batteries is this: Do not charge them unattended. Lithium ion batteries are very dangerous (think about what recently happened with the Samsung Galaxy 7 Note) and never more so when they are charging. I always place mine inside a metal tray on a kitchen counter or on a concrete garage floor and charge them there. Lithium ion cell fires are lethal in nature and short of a short list of very expensive fire extinguishing agents, cannot be extinguished. Keep alert. Many drone pilots keep them and transport them in special lithium ion battery bags. Oh, and never leave batteries in a hot car!

Other maintenance items include ensuring that your drone software is up to date and your controller software is also up to date. Your user's manual should outline the procedure to updating your software. DJI and any others that use smartphone and/or tablet apps also require you to update your app as needed.

Finally, and this is extremely important. Make sure that you calibrate your compass any time you have travelled 20 miles away from where it was last calibrated. It is not a bad idea to do so every time you start to fly in a new location. Be very careful when calibrating it. Do not calibrate it near a large metal structure (or concrete structure with a lot of metal rebar) or near to or under power lines. Be also aware that some locations have underground power lines so that can contribute to distorting the Earth's magnetic field. After you first take off, check that the drone flies in a stable fashion. If it does not, recalibrate the compass in a different area.

Rentals: Try Before You Buy

This is targeted at anyone who is looking to purchase a very expensive drone such as a DJI Inspire or Matrice. Before you commit to such an expenditure, you may need to try it out and there are a growing number of movie equipment rental houses that carry drones. Many even carry the less expensive models such as the DJI Mavic and Phantom. Each rental house has its own requirements for deposits and insurance so you may need to shop around a bit.

Rental houses also provide excellent options to try specialty cameras such as the FLIR infrared camera. New, these can run upward of $7K, so to try it out and even do some training on it, a rental may be in order.

Software Automation

Many, if not all higher-end drones, have applications to drive the flight and maintenance processes.   DJI drones can also take advantage of 3rd party apps such as Litchi, Altizure, Vertical Studio, and DroneDeploy.

All of these applications bring advanced automation to the drone operator's fingertips. Some features that they open up include Orbiting, rotating around a defined point for filming moves; Waypoint Tracking, marking a series of GPS points and altitudes and following them repeatedly on different missions; Focus Tracking, which permits a camera to be locked to a specific point and follow that point no matter where the drone goes; Tracking, which is flying in a straight line from one point to another regardless of the orientation of the drone's front; and 3-D Mapping, which surveys an area to create a three-dimensional map.

Familiarize yourself with the software that comes with your drone, how to enter into these automation modes and how to exit them, and also check out what 3rd party applications exist for your model. Automation can help save the day on a particularly difficult shooting scenario.

Manual Mode: Your Friend in Need

All of the automated modes that live inside the application software and the GPS advanced positioning capabilities make for a tremendous flight experience 99% of the time. It is important to always be vigilant while flying because even the most routine of flights can suddenly degrade due to unforeseen circumstances. The more experience and practice that you have under your belt will make things less scary.

One of the greatest allies you have when things go crazy is actually being able to take absolute control of the aircraft. This may seem counterintuitive, but consider that if a drone "decides" to do something unexpected it may be related to one of the automation processes running amok. This could occur due to several reasons such as a loss of GPS signal, a misreading of a compass position, a miscalibration or failure of some component, or an actual software glitch. Many drone "fly-aways" or crashes happen when these situations occur.

Taking full control of the drone is best in most, if not all, of these situations. To do so normally involves clicking a switch on the controller of most drones. It is important to know how to flip into Manual or ATTI (Attitude) mode. Once this has been engaged, the automation drops offline and you have control of the flight using your control sticks. If the drone was in the process of a "fly-away" scenario, it should stop dead in its tracks and permit you to control its return and landing.

Of course, there is no free lunch. Manual mode is your friend but you must also become comfortable with the process of flying without the GPS safety net. To become comfortable flying this way, you should build-in practice sessions at safe locations as described in earlier sections. The more you practice, the more you will be ready to deal with the unexpected.

Practice Maneuvers

As we wrap up this section, it should be clear that the more that you practice flight with your drone the better you will be able to deal with normal and abnormal situations. It is advisable to take time to practice maneuvers in a safe flying location as often as practically possible. Choose a location away from airports and without a large number of hazards and take time to familiarize yourself with the controls, the software, and the characteristics of the drone. In my area, I have found a number of city- and county-owned softball and soccer fields which are remarkably free of people a lot of the time and I tend to use them as my personal practice fields.

Some example exercises were suggested in earlier segments. Feel free to make up any number of challenges and try them out. If your drone usage is mostly centered on filming, practice moves such as reveals, tracks, and orbits. The point is to practice with and without automation so that you will become a better drone operator. Keep track of your practice time also in your logs and if you learned something very significant, you might want to note it in the log also for future reference.

One more thing, fly in as many conditions as legally possible. Flying in fog or on extremely gusty days is not permitted by the FAA, but there is nothing preventing you from flying on a cold day or a very hot day. Gaining experience in how the battery life and lift characteristics change in these conditions can be helpful when you have to plan a more critical mission in one of these extremes. You will know how to keep batteries warm in winter, how much less flight time you will get when the props have to work extra hard in the high-density altitude conditions of a scorching summer day, and so on.

If you haven't gotten the message yet, the point is to just keep practicing. Practice makes perfect.

Learning to Fly Fixed-Wing Drones

Fixed-Wing Drones

When one says "drone" most people immediately think about multi-rotor craft. While they certainly have captured the bulk of the market, they are not the only types of drone. The first ever use of the word "drone" was during World War II and it referred to a radio-controlled fixed-wing aircraft produced by the Radioplane Company for the US Navy and which was used to train anti-aircraft gunnery skills.

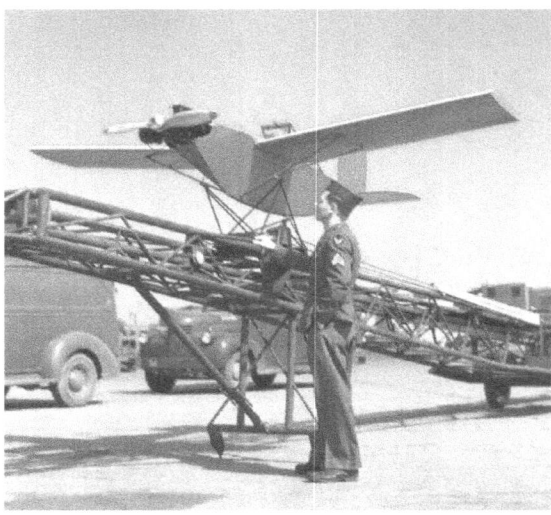

Figure 11 OQ-3 Radioplane "Drone" (photo Wikipedia)

Ironically, in some circles, the word "drone" is returning to this original use.

What is the appeal of fixed-wing drones? First of all, they are far more efficient per ounce than their rotorcraft cousins. This is because they do not need to waste energy in generating lift. Since the lift that they need is mostly generated across their wing surfaces, the only energy that is expended is that in maintaining forward motion. This does enhance a liability of fixed-wing drones versus rotorcraft though: they must always be moving forwards to keep aloft. This observation leads to the second reason to choose fixed-wing over rotorcraft in certain applications. When the application involves surveying large swaths of land (mapping and surveying, agriculture, real-estate video of large farms, large scale search-and-rescue, forestry applications, etc.) which involve long parallel survey tracks, the fixed-wing is in a world of its own. In some applications such as S&R

(Search and Rescue) or military/police surveillance of fence lines, a combination of craft may actually be called for -- a fixed-wing to do the large-scale survey and a rotorcraft to fly in and apply its precision operation to a hot-spot.

Fixed-wing manufacturers' offerings range from hobbyist and dual-use drones (such as KKmoon, Parrot, Hubsan, UAV Systems) to expensive commercial and military use drones (such as Trimble, Autel, Saxon, Insitu, Airborne Concepts) that can cost thousands through millions of dollars! An interesting development in this arena are the hybrid VTOL (vertical take-off and landing) drones, fixed-wing drones that behave similarly to the Osprey (e.g. the Kestrel). They either have rotating props (like the Osprey) that point up for takeoff and then rotate to horizontal for flight or they have props that assist takeoff and landing but which feather during forward flight.

Fixed-Wings are Different

While some fixed-wing drones have impressive control software that can make a rank beginner seem to know what they are doing, should something go horribly wrong it is best to know how to fly. So, do not make the mistake of thinking that since you can fly rotorcraft drones that you can just purchase and fly a fixed-wing drone out of the box! The two are totally different and skills learned flying one do not necessarily translate to the other (although, I have been told by some that flying fixed-wings will translate back to rotorcraft in the form of smoother operation of the control sticks).

The best first step in preparing to do fixed-wing drone flight is in learning to fly RC (radio controlled) aircraft. In several respects, fixed-wings are more complex flight systems than multirotor craft. The first of these is in the realm of flight itself. Rotorcraft have intricate software which convert the intent of the pilot at the controls into actual flight data. We may drive the controls to ascend to a specific altitude and move forwards slowly while rotating clockwise and the craft's "brain" converts these inputs into the proper prop RPMs to make these things happen. On a fixed-wing system, however, lift is generated across the wing so gaining altitude and positioning the aircraft requires piloting.

The second way in which fixed-wings are more complex to pilot is that one must understand and grasp more than just the basics of aerodynamics. Since lift is controlled across the surfaces of the aircraft there is a complex interaction between the thrust of the propeller, the attitude (angle of attack) of the aircraft, the basic glide and self-righting characteristics of the aircraft, the effects of the direction and intensity of winds, and the positions of the control surfaces. The understanding of this comes down to a combination of knowledge and flight experience. Hours of experience count towards any drone flight, but most especially to fixed-wing systems.

A third difference is found in the fact that most fixed-wing drones require a decent amount of space to land. Even automated landing systems require a long, clear path to accommodate their approach pattern and landing space. A large number of these drones come in on their bellies so they need the landing area to be grassy and relatively flat. This becomes especially important while operating in the commercial drone space where one must select and plan for alternate landing areas in case of emergencies. A multirotor can set down almost anywhere whereas a fixed-wing drone isn't as forgiving.

There are a plethora of flying parks across the US and in many other countries. The best place to start in the journey towards RC fixed-wing flight is to join up with one of them. The Academy of Model Aeronautics (AMA) has a site to search for affiliated clubs within the US at their search site found at http://www.modelaircraft.org/clubsearch.aspx. As mentioned elsewhere in this book, it behooves the drone pilot to become a member of this organization since it advocates for model aircraft use (which includes drones) and offers a great liability insurance policy for non-commercial use of model aircraft. In fact, many clubs require their members to be commensurately members of the AMA to ensure proper insurance coverage.

Another good link to find links to clubs including in international locations is on R/C Airplane World's site http://www.rc-airplane-world.com/rc-airplane-clubs.html.

"I'm Learning to Fly"

My wife and I both started on the journey to fixed-wing and I will share the process here in case you are interested in following this path at some time. The first thing is to get a trainer aircraft. We chose the Flyzone Sensei and were also eyeballing the E-flight Timber. Good aircraft to learn flight techniques are larger than you would expect because beginners need to be able to spot them easily. Good trainers also have a high-wing design because this setup is more self-stabilizing and makes mistakes easier to deal with.

Trainers can be generally bought in Ready-To-Fly (RTF) kits which include the remote control, or in Receiver-Ready (RxR) kits which contain wiring and servos but the customer must provide the controller and receiver. Some assembly and testing is generally required to get these kits prepared. If you are already proficient at assembly and electrical wiring, you can do this yourself and have an experienced RC aircraft person check your work. If you are not, then coordinate with your RC club to see if someone can assist you in this phase.

Once assembled, the first thing to do would be to range-check your radio. This ensures that your receiver and radio are communicating well at all angles. Your training can now begin with exercises to learn how to taxi your aircraft. Practicing how to maintain a straight line and to do circular and figure-8 patterns while the aircraft is on the ground will begin the ever-important process of learning how to use the throttle and the rudder and nose/tail wheel delicately on the left stick (assuming standard mode-2 setup of the controller). Make sure that you start with slow movement and build up as you become more attuned to the controls. Also, if you find you are moving too fast and the aircraft wants to lift off, do not panic. Merely cut back the throttle and let it slow down and settle back on the ground.

The next steps are best done with a coach at an air park. This experienced flier can check the airworthiness of the aircraft, fly it around a bit to test it, and then set up a "buddy box" connection between your controller and theirs. This way they can keep control of the aircraft and explicitly hand over control for you to learn how to do a maneuver. Should anything go wrong, they can immediately resume control and correct. Bear in mind that you may want to locate and speak with your coach before ordering your aircraft since different brands of controllers use different buddy-box connections. You will need to ensure you have the same brand as your coach's or check what brands they can cross-connect to.

The coach will start working you through understanding the basics of aerodynamics. If you are used to multi-rotor GPS-stabilized drones, you will quickly realize how much the software in the drones actually does for you! For fixed-wing flight to occur, lift must be generated by the wings and this is accomplished by using the throttle to create forward movement. An exception to this rule exists, of course, in which the lift can be created by the prop if there is enough power to go vertical and hover the craft, but this is beyond mere flight as it is in the realm of acrobatic flight!

So, there must be thrust from the propeller that is sufficient to create an appropriate airflow over the wings and control surfaces of the aircraft in order to create lift. Adding more lift by increasing speed or using up-elevators to change the wings' angle of attack will cause the aircraft to climb. Too much angle of attack or too little power will disturb the airflow over the wings and lead to what is called a stall. A stall condition leads the aircraft to lose altitude and needs to be corrected by increasing power and/or decreasing the wings' angle of attack (by moving the elevators downwards). The increased speed or improved airflow over the wings will generate lift again. Turning the aircraft can either be done through activation of the ailerons to tilt the wings into the turn and simultaneously adding some elevator lift to compensate for the plane's desire to drop its nose, or in some aircraft which do not support ailerons this can be done by using the rudder.

Reversed Controls

Remember when we discussed learning to fly or the multirotor drones that one of the problems that had to be overcome was the fact that when the drone faced in any direction other than away from the pilot, the controls changed. Specifically, the controls on the right stick were affected. Forwards/backwards and left/right on the drone no longer corresponded to these directions relative to the pilot. Thus, it was recommended to practice slowly with a toy quad orienting the craft facing 90 degrees, 180 degrees, and 270 degrees relative to the pilot and learning how to move pilot-relative left/right and forward/backward.

Well, the same kind of control changes happen with fixed-wing aircraft, except it affects the side-to-side controls on both the left and right sticks (running in Mode-2 on the controller). The left stick side-to-side controls the rudder (which points the nose of the aircraft left and right around the yaw axis) and the right stick side-to-side motion controls the ailerons (which move the aircraft left and right by rotating the wings around the roll axis). Either of these (in the case of simple controllers without at least 4 channels), or both of these (in the case of fully wired controllers and aircraft), will be reversed when the aircraft is heading towards the pilot. In fact, one of the major reasons for beginners crashing is because of this phenomenon. Overcoming it can be done through practice with an experienced trainer pilot and definitely by using a flight simulator in pilot-view (ground) mode

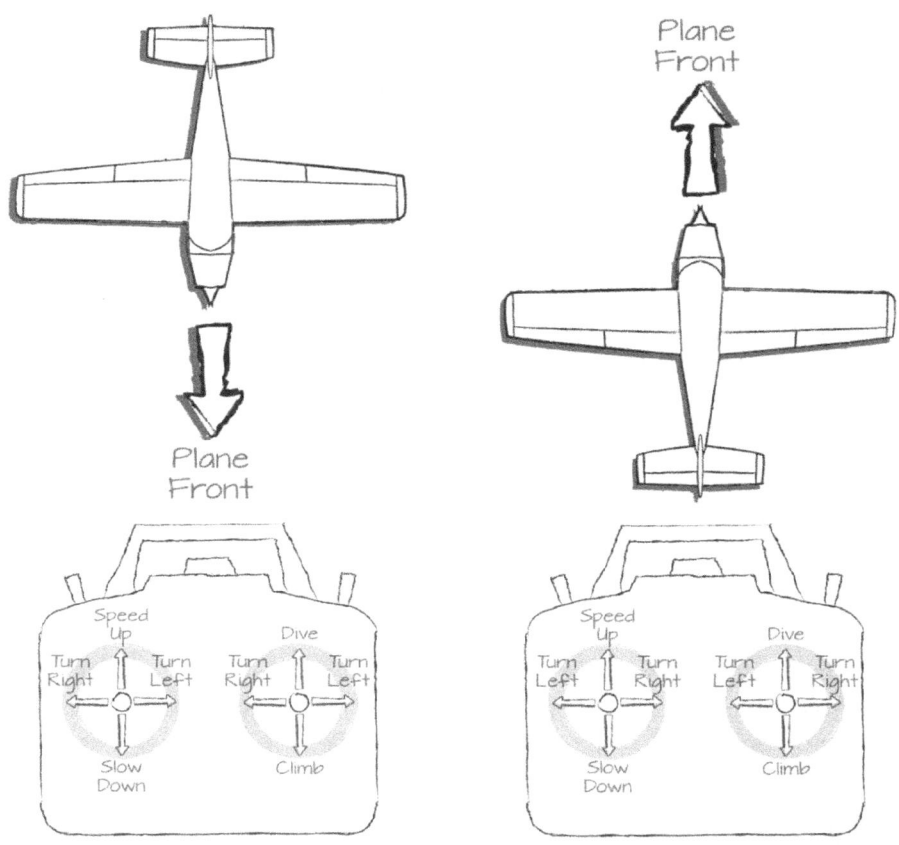

Figure 12 Reversed controls relative to the pilot when the aircraft changes direction

.

R/C Flight Simulators

If you have the money and an appropriate computer, purchasing a good simulator that either interfaces to your RC controller, or one that comes with its own (e.g. in my case, RealFlight eXperience (RF-X) with the InterLink-X controller) is a good idea. Most simulators allow you to fly both fixed-wing airplanes and rotorcraft drones so they can help you hone all your skills. If you only fly rotorcraft, there are simulators only equipped for such drones.

The beauty of the simulator is that it can serve to improve your controller skills (smoothness and precision) in a way that doesn't damage your actual aircraft. The most important thing to grasp when using simulators is to not fly crazily on them. The habits you develop playing on the simulator hour after hour will manifest themselves when you actually fly your real plane, and "fun" crashes on the sim will turn into expensive crashes on the field!

How do Fixed-Wing R/C Planes and Fixed-Wing Drones Differ?

It is important at this juncture to stress that a traditional fixed-wing R/C airplane is not a drone. They normally do not use GPS positioning algorithms to drive their flight. They do not normally provide a FPV (first-person view) camera (although this can be added to R/C planes as an add-on). They normally do not provide comprehensive instrumentation data back to the pilot (but, again, this can sometimes be bolted on to an R/C plane). Finally, their flight time tends to be short in terms of drone flight times because they tend to be flown close in and for short periods of time. Face it, there is only so much concentration that a person can have doing aerobatics before they need a mental rest.

Drones, on the other hand, tend to have all of the above systems and more. Even the cheaper fixed-wing drones provide some sort of GPS automation, FPV camera data broadcasting back to the pilot's control console, and instrumentation indicating direction, altitude, and remaining battery power. Drones are made to remain aloft for longer periods of time, measured from as little as ½ hour out to hours! Drones, unlike R/C planes are made to be flown in applications that are far enough out from their pilots that instrumentation and GPS support becomes critical.

Consider, for example, providing knowledge back about battery power is a major difference. Most R/C pilots use a simple timer to inform them when their battery or fuel should be running low. On my trainer, my timer is set to 6 minutes because we know that 8 minutes is too long (by trial and error)! This would never cut it for a drone application because it is too short a time to be aloft and guessing is not a good plan when one has thousands of dollars of equipment flying overhead!

Why Not Begin by Buying a GPS Fixed-Wing Drone?

There are tons of videos online for GPS and electronically stabilized drones on the Web. They make flight seem mindless and are extremely tempting to make you jump straight into them. However, as you just saw, fixed-wing flight is very nuanced. Working the scenarios on a trainer with a good coach is the best way to learn without crashing your plane. Over time you will develop an instinctive muscle memory for how to deal with different flight scenarios. This is similar to the muscle memory that you would develop on a toy (unstabilized) quadcopter before flying a full GPS-driven quad.

A great example of why this is important is the Parrot Disco FPV Official Video (search for it on YouTube). It is pretty exciting. Just throw it up into the air and you too can be a fixed-wing drone pilot! Oh, I am not picking on Parrot, by the way. Personally, I think that the Disco is a pretty interesting drone and with its recent price drop, it could be a good entry path into my own fixed-wing drone exercises sometime in the future. The point I want to stress is that there are those who may want to skip learning the basics and get the drone itself and falsely assume that the drone has *all* the smarts to always keep them out of trouble.

A quick Web search on "Parrot Disco Crash" will hopefully demonstrate that even seasoned fixed-wing fliers can crash this technology-marvel. A good article with accompanying video can be found on Quartz (https://qz.com/767788/this-is-what-it-looks-like-to-crash-a-brand-new-1300-plane-like-drone/). Notice again that this crash was not due the fault of the aircraft itself but it was operator error. Just like their rotor-equipped cousins, fixed-wing drones depend on the operator knowing how to control them should the eventuality arise.

As mentioned earlier, joining a flying club in your area is the best approach for learning how to fly. Visit the club's field ahead of time, if at all possible, before you purchase a plane. Try to meet someone who is available to teach you. They will recommend a plane to purchase and the controller that they can connect to with their buddy-box cable. Jumping into fixed wing flight will cost about $150-$400 depending upon the choices you make.

It is probably a good conservative estimate that it will take 3 to 6 months of relatively regular sessions each weekend before you will be to the point of knowing how to fly well enough to really know what to look for in a drone. There is a whole new language to fixed wing flight and learning to visualize the aircraft's movement in 3D space is something that takes time. Just as in the case of their rotorcraft cousins, they may have great technology to help stabilize and orient them but it is important for a pilot to be able to take control when necessary. Ultimately it is better to learn how to avoid problems by crashing an inexpensive aircraft on a field than an expensive drone with an equally expensive camera rig into someone's building.

Filmmaking Moves

A Few Points about Camera Moves

While filming, moving or panning a camera left and right, zooming in and out, and moving up and down may seem really cool to a beginning videographer. However, moving a camera around needlessly actually detracts from the flow of images that you are producing with your camera. Think about what happens when a rank amateur first gets their hands on a camcorder. They zoom in and out, whip pans side to side, and produce something that gives the audience motion sickness.

Consider the camera work on top-notch movies. Movements of the camera are actually minimized and tend to flow with the story. Even a conversation between two people is not shot by whipping the camera back and forth between each person's face. It is generally shot from two or more angles and then the back-and-forth banter is cut together. Moves of the camera in 3-D space tend to be very deliberate and motivated.

Shooting aerial footage with a sUAS that can hover like a helicopter and then move freely through 3-D space may tempt you to shoot motion all the time. The question that you will need to ask yourself is if motion is used, what type of motion should it be (pan, push, pull, jib), and how fast or slow does it need to be. If the shot is a chase scene that is meant to emulate the point-of-view of a police helicopter, then it will be completely different in its characteristics from a slow jib reveal climbing up a tower to show a town behind it. Some filming may well require sweeping motion while others might be shot with the drone perfectly still. Consider that you may frame up a scene in a park and hold it while two actors approach each other and initiate a clandestine conversation. The aerial footage may then cut to ground-level footage of the two talking to each other.

The mass of a drone makes it very jerky when starting and stopping moves unless you are very careful with the controls. Many controllers permit the operator to extend the control sticks by unscrewing the tips and lengthening the sticks themselves. This serves to give more control since a longer radius for the control lever means that a longer throw is needed to get the same change in angle. A shorter throw results in a smaller angle which translates to a slower and smoother transition. Some drone controllers also permit the operator to reprogram the sensitivity of the controls to make them less instantaneous in their responses. It goes without saying that if you do change the sensitivity of your controls, you need to tread very carefully and make very small incremental changes followed by extensive testing before settling in on a setting.

Any controls that move the camera's gimbal need to be set to move slowly or reasonably slowly. The objective should always be to ensure that starting and stopping a gimbal move is gradual. If possible, program pitch and any other gimbal controls to use an exponential or bezier curve rather than a linear response. Over the years, experienced Hollywood camera operators perfected tilts and pans on their tripods by setting the controls to the right amount of drag and then using a large rubber-band to pull the handle on the tripod head instead of using their fingers directly. This served to introduce a certain amount of exponential ease-in and ease-out to the movement and reduced the possibility of an inadvertent jerk on the movement. Anything that can be done to the controls for the camera gimbal to smoothen out its movement will result in much better filming.

Flying a camera on a fixed-wing drone serves a different purpose. Most fixed-wing drones are mostly equipped for downward viewing applications such as surveying and crop monitoring but others carry a front-mounted camera. Gimbaled or not, the footage from this camera is mostly used to emulate the view from an aircraft's cockpit. The best way to create footage of a dogfight between military aircraft (think "Top Gun" for example) on a limited budget is to use fixed-wing R/C aircraft or drones. The footage will be more believable for this application than trying to fake it with a multirotor drone!

The Importance of Reveals

One of the primary cinematic moves, one which is extremely familiar and yet still absolutely effective, is called the reveal. On the ground, one way a reveal works by using a dolly or something similar which tracks a subject as it moves. As the subject is tracked and the dolly follows the action, it suddenly reveals another subject of interest. Perhaps, around the corner that the original subject is approaching, the reveal shows an axe murder poised to deliver a death blow. Another use of the reveal is to expose some subject of interest. In this example, a dolly shot, a helicopter, or a crane or jib may track over or around some trees or foreground object to suddenly expose a stunning view of a cityscape. This creates a sense of awe or expectancy in the audience.

Drones can also pull off very effective reveals. These can be orchestrated with sideways dolly-like movements, vertical jib rises over some foreground object, or by using a combination of sideways, vertical, and forward/backward moves. There exists very little that compares to a buttery smooth filmic reveal transition for setting up a scene. The key to this move is two-fold.

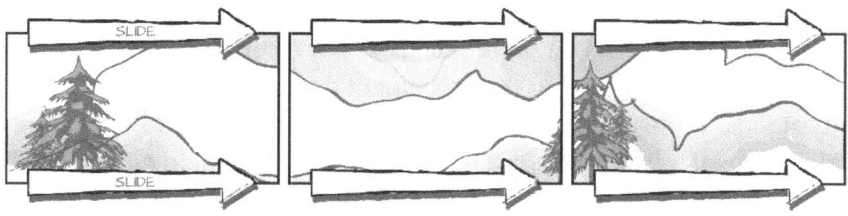

Figure 13 Example of a slide move

First, it must be very smooth and well-planned out. A jerky slide will distract horribly from the effect. To keep it smooth, it is best to begin the move before the IN point for the edit and, if possible and the script does not call for dwelling on the subject, carry the move forward beyond the OUT point for the edit before stopping. If the script requires a stop on either end, then practice a smooth and consistent acceleration or deceleration so that the movement is completely smooth. This requires really knowing your drone's characteristics and being able to properly control its transition of mass.

The second key to the reveal is to make it interesting. Do not just do a reveal for a reveal's sake. As with all filmic transitions, a move like this should serve the storyline. Reveals work well for transitioning into a new scene, and sometimes to transition out of one. Reveals also work for suspenseful purposes. Regardless of the actual motivation, the reveal should have a defined start point (the IN point) and a defined end (the OUT point). The move must be practiced until it is right. The camera needs to be properly controlled to not be out of focus at either extreme, and be very careful if using auto-exposure modes because while the move may be smooth, if there is an abrupt change of aperture, that will be perceived as jolting to the audience.

Push Ins, Pull Outs, Trucks, and Pans

Imagine a subject framed by the camera which then slowly moves closer to the subject and makes them command more of the area of the frame. This is known as a push-in and is a common cinematic move that allows the audience to get a better view of the subject or emphasizes the subject by making them more dominant. Some push-ins are slow and deliberate, maybe as some dialog is being recited, and others are quicker and more emphatic.

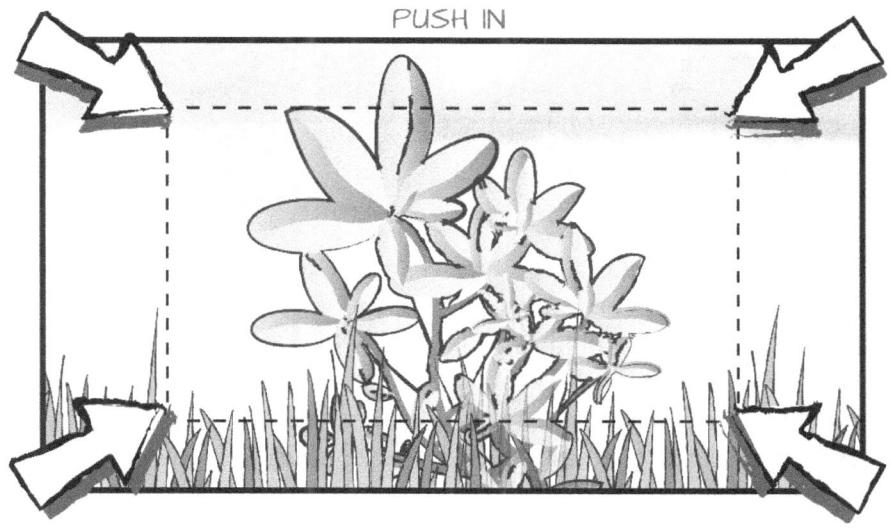

Figure 14 Example of a push-in

A pull-out (or pull-away) is the opposite of a push-in. The camera pulls back from the subject and generally permits the audience to capture a feel of the environment surrounding the character. Imagine a tightly-framed image of a knight with drawn sword reciting a line about how he will not be overcome. A pull-out may then occur to see him surrounded by a dozen other well-armed knights all focused on our hero. The brave line falls flat as you realize that this situation is impossible and the hero is about to face a desperate battle at the least, and maybe his own demise at the hands of so many enemies.

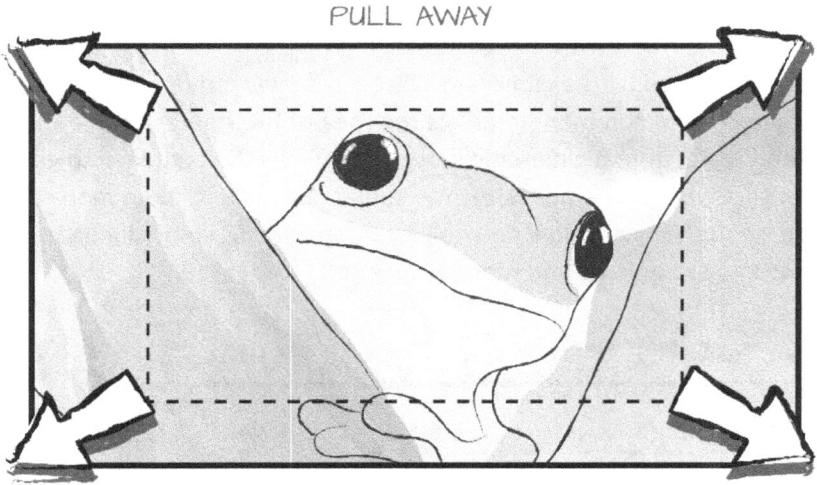

PULL AWAY

Figure 15 Example of a pull-out/pull-away

It is important to understand that push-ins and pull-outs are not zooms. Moviemakers rarely use zooming actions because zooms tend to change perspective and relationships within a scene. A push-in or pull-out achieves the change in size of the subject by physically moving the camera closer or farther. In countless movies, these moves were made with a camera on a tripod mounted to a set of rails. This setup is referred to a dolly and thus push-ins, pull-outs, and sideways slides are all referred to as "dolly shots."

While on the subject of zoom lenses, rarely these are found on today's drones and when they are, they rarely permit zooming while actually filming. Zoom lenses on drones tend to permit the zoom to be activated to frame up a shot then the zoom is locked out while filming is occurring.

Sideways slides left or right are called "trucking shots." Again, in the moviemaking world, these were accomplished by laying down track parallel to the subject and the entire camera moved sideways while remaining locked facing 90 degrees to the motion. Trucking shots can be used to expose an entire expanse of a building or to follow a subject who is walking or running. Notice that a trucking or slide shot is not the same as a pan left or right.

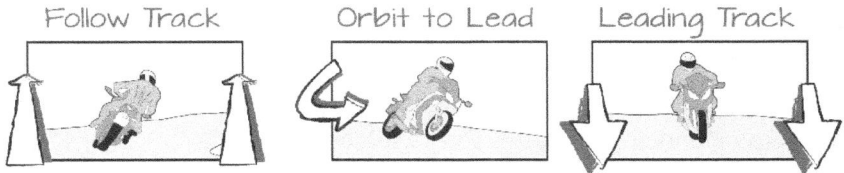

Figure 16 Example of a trucking or slide shot.

Push-in, pull-out, and trucking shots require that the drone move while the camera gimbal remains locked in place. Since these moves involve movement of the heavy drone, they will need to be practiced to make them smooth. It is easy for these movements to become jerky whenever such a large mass is being moved. The suggestions in the previous section about smoothening out the controls may be helpful and continual practice will improve your feel of the controls. Be also aware that filming on gusty days will make it even more difficult to be smooth!

Pan shots occur when the camera is rotated left or right centered on an axis that traverses the camera. Most amateur film uses pans when the camera operator needs to show something moving left or right. To do this move, the drone is held in one place and the gimbal is rotated to the left or the right while filming. An alternative is to actually keep the camera gimbal locked off and rotate the entire drone. This is actually the only way a pan can be achieved with a 2-D gimbal rig. Of course, if your rig has a 3-D gimbal, it is smoother to move the gimbal instead of the heavy drone!

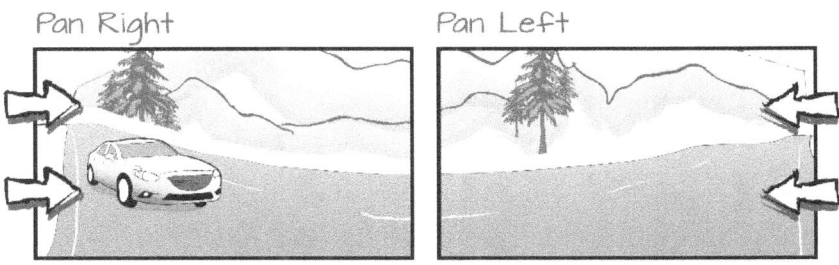

Figure 17 Example of a pan movement

75

Jib Movements

So far, we have described movements that occur in a 2-D plane. Of course, a drone can move freely throughout the 3-D Dronespace so it can easily achieve motions that would normally require large and heavy rigs. These movements move the camera up and down relative to the subject.

In the film-making world, a short movement up or down (a matter of a few feet) used be achieved by a mechanism on the camera called a pedestal. Thus it is called a pedestal up or pedestal down movement. Larger movements are normally handled using a crane and are called jib or crane shots.

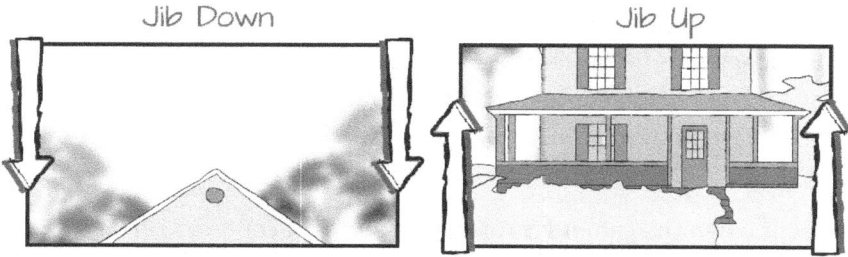

Figure 18 Example of a jib shot

These shots can either remain centered on a subject or rise above the subject. The difference depends upon if the camera is tilted (pitched) up or down to keep the subject in the frame while the drone moves down or up or not. A jib up is a difficult move to get smooth since it involves adding lift to an already heavy drone. Practice on how much power needs to be added incrementally to start the move and then how much more is needed to continue feeding the motion in a smooth and consistent way. At the upper end of the move, don't just chop the power but taper it off until the drone is hovering again. If the camera is meant to pitch downwards at the same time, you more than likely will need a second person controlling this for you.

Orbits and Yaws

We have all seen lovely helicopter footage where the camera floats around the subject of the scene or rotates slowly to show some wonderful panorama. Drones

76

are able to produce the same kind of footage. I many cases, the software on the controller even can assist in creating this effect.

An orbit occurs when the drone is centered on a point and flies around that point at some speed. DJI and other manufacturers' along with 3rd party software have modes that enable orbiting. If your drone does not have this capability, you can practice this by simultaneously moving the right control to the left or right and simultaneously adding a slight amount of the opposite motion on the left stick to yaw right or left. This must be a well-coordinated motion to keep the subject in frame.

Orbit

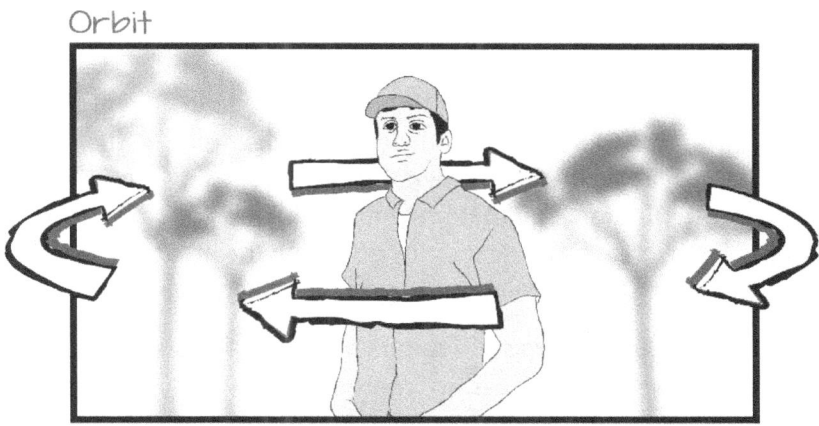

Figure 19 Example of an orbit

Luckily, with the software option, all of these calculations can be done by the controller software and all you need to do is to set the center of interest (normally by flying over it and marking the spot), and then moving the drone out as far as you need it to be, setting its desired altitude, focusing and starting your camera, and then setting an orbit speed. While orbiting, you can normally change the speed of the orbit, altitude, and even move the camera to point elsewhere. Check the documentation for your drone to fully understand what you can do in orbit mode.

A drone's yaw motion is actually the same thing as a pan. A common creative use of a yaw is to yaw left or right with the camera pointing to the outside of the circle to get a motion similar to looking outwards from a carousel.

Tracking Shots and Waypoints

Sometimes it may make sense to lay out a path for a drone to follow so that repeatable shot sequences can be captured. One of the uses for this is to be able to follow the same track for some inspection routine that can be executed as a schedule. Filmmakers can also take advantage of this capability to lay in a complex path for a filming track and be able to repeat it for take after take.

How this can be accomplished is by using waypoints on drone software that support them. Some implementations of this require you to fly the mission and mark waypoints from the GPS position and altitude of the drone. Other implementations such as Litchi permit waypoints to be set using a map to set GPS points and altitudes. DJI's implementation requires the mission to be flown and, in my humble opinion, creates a limitation by enforcing the minimum altitude of the waypoint at 33 feet. This certainly can make it less than ideal for a filmmaker.

Once the waypoints are set, the mission can be saved and replayed as often as needed. Generally on replay, the operator can specify how to fly the mission. One option is to fly it exactly as recorded while other options permit the operator to rotate the drone and/or camera dynamically, execute turns by flying to the waypoint and turning or by curving smoothly around the waypoint while flying, and so on. As with all automation, practice in a safe location needs to be done to understand how it works on your specific drone.

Tracking shots can be done manually by using a slide/trucking approach as described earlier. However, sometimes it would make more sense to use automation to control this shot so that you can then do creative moves with your drone while not running the risk of deviating from the track. Most of the automation programs permit setting two or more waypoints and then precisely running that course while controlling the speed and orientation of the drone's front and/or camera.

Imagine a scene where the drone needs to run parallel to a motorcycle speeding down a city street. Flying manually, there is the risk of the drone colliding with the rider, especially if some sort of pan maneuver is needed. Setting a track and locking the drone to run that track frees the operator from needing to concentrate on following the path and lets them concentrate exclusively on controlling the speed and rotation of the drone. As long as the motorcycle rider follows their assigned track, there is no risk of collision.

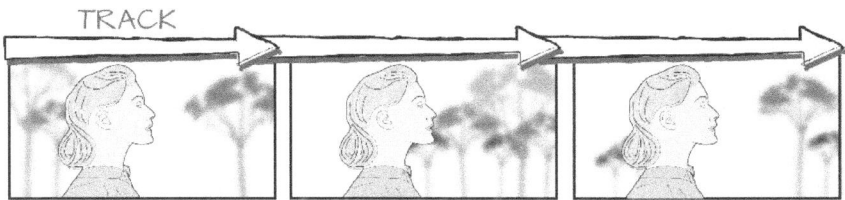

Figure 20 Example of a tracking shot

Another type of tracking mode is also available in some software packages which is called target tracking or follow tracking. Using this mode, the operator can click on a target or draw a box around a target then start recording. The drone will use pattern recognition to keep tracking the target as it moves. As with all target tracking algorithms, it can get fooled by changes in the background or the aspect of the target so do test it carefully, be fully aware of obstacles in the environment, and be always ready to assume control of the drone if something goes wrong!

Follow-Me Tracking

The final filmmaking move that we will touch on is the "follow me" tracking mode that is available on several drones. Some sUAS controllers actually have GPS locators which make this mode even more accurate. Others use triangulation methods to sense where the controller is relative to the drone itself. Yet others, like Litchi, permit a phone to download and use a tethering app which relays GPS coordinates back to the controller. The basic operation of this mode is that the drone will follow the controller or tethered phone and point the camera at its coordinates.

Bear in mind that this mode is potentially fraught with danger. I say this, not to dissuade you from using it, but so that you will be aware of how you use it. If the target is a person walking, then sure you can use it in most locations. However, if your goal is to chase a car, I urge you to not attempt using this mode anywhere other than a sparsely populated rural area. If you think about it a moment, there are so many things that can go wrong with it that you need to be extremely careful to minimize each source of risk.

On the other hand, used properly, this mode can really lend itself to a foot or vehicular chase scene. Do bear in mind that many (if not all) software packages that offer this limit the top speed for tracking. DJI has a limit in the 20-25 mph range and Litchi's is in the 30-35 mph range.

Some drones have anti-collision optical sensors (for example, the Phantom 4 or Mavic) so they can sense obstacles ahead of them and avoid them. Others don't have these systems. So, it is critical that you understand all of the aspects of the path your subject is going to follow so that you don't run into tree limbs, overpasses, powerlines, and other such obstructions while using this mode. You have the option to set the altitude that the drone will keep above the subject, so if there are obstructions, do ensure that your minimum altitude will clear them.

Again, use this mode carefully and always be ready to assume control if anything goes wrong. Finally, remember that the FAA requires that drone usage is always within line-of-sight of the operator or by a radio-connected visual observer, so while using this mode make sure that you will meet this requirement.

Making Money: Specialized Uses for Drones

Real-Estate Photography and Videography

One of the hottest emerging markets for sUAS technology at the time of writing this book is the use of drones to film and to photograph real-estate. This is partially because the dramatic aerial angles of property may attract potential buyers and is partially because of the ability for an aerial image to help a buyer visualize where the property sits in the landscape.

Using filmmaking moves such as reveals and orbits can make real-estate videography more dramatic and exciting. The ability to fly over a property and do a 360 degree view can help developers gain a good feel for the visibility of that property for signage and how adjacent properties would fit into or hinder their proposed plans. Overflight videos of large properties such as estates can make them more accessible to buyers.

High-quality aerial photography can also enhance a property. Sometimes the most favorable view of a house is not at ground level. Merely being able to rise 10 feet may make for a better photograph and a drone can easily do this. Other photos that are in demand are angled shots that show how a house fits into its environment. Finally, some real-estate agents want a good high-quality Google Earth-like straight down shot that outlines the shape of the building. Of course, remember that the FAA's maximum permitted height for the drone is 400 feet above the structure so don't think you can pop into lower-Earth orbit for this kind of photograph!

Additionally, developers and construction companies also use drone video and photos to track a timeline for their projects. In fact, sometimes these are required by their financiers to prove that their projects are meeting required timeline milestones.

Since this is a commercial endeavor, filming and photographing real-estate must be handled by a licensed pilot-in-charge. Some construction and real-estate locations are within space dominated by an airport so the PIC needs to coordinate with the airport manager and/or the NAS (National Airspace System) to secure permission to fly, file NOTAMs, and ensure that the mission goes as planned. Additionally, most real-estate operations require that the drone operator be insured.

Insurance Documentation

Another interesting emerging market for drone use is filming and photographing locations for insurance claims. This may be a simple as a single property that has suffered some sort of damage that occurred or it may involve documenting a large number of properties that may have been at the epicenter of a national disaster. The ability to document the damage as soon as possible after a catastrophic event can help speed reimbursements. Drones are able to easily cross over police lines and flooded areas and gather views that might not otherwise be accessible for some time.

Equipment, Building, and Tower Inspection

A really hot market for drone pilots is the inspection business. Inspection of power lines, trains, oil and gas rigs, and building structures normally are time-consuming and potentially hazardous for individuals. Inspection of power pylons and towers by tower-climbers is inherently risky. However, using drones to conduct these inspections can help increase both the productivity and the safety of inspectors. Only when an issue is discovered that needs direct intervention will a person actually have to climb a tower or work their way through a busy railyard.

A few things to consider when it comes to the inspection of towers and buildings is that the FAA permits a drone to fly 400 feet above a structure while within a radius of 400 feet of that structure. This translates to the fact that if a tower is 1,000 feet tall, you are not restricted to 400 feet above the ground while flying within 400 feet of that structure. Your ceiling is 1,400 feet AGL (Above ground level) over that tower which permits you to do a complete visual inspection of that structure. If that structure is a power pylon or transmission tower, you need to be aware that it may interfere with the compass and electronics of your drone. Be always ready to flip the drone into a manual mode and take control of a runaway situation.

Another thing to be aware of when inspecting towers and pylons is to be aware and avoid guy-wires and/or electrical wires. These have a habit of disappearing from view against a bright skyline so be extremely observant as you maneuver or your drone may collide with a wire and crash.

Infrared and Thermal Cameras

Finally, another emerging technology for commercial drone operations is in thermal scanning. Companies such as FLIR have developed miniaturized infrared cameras for drones which can be used to check structures for energy loss or for search-and-rescue teams to use to search for missing people. These cameras may not be cheap but they are far cheaper than a full helicopter equipped with a full-sized unit!

A very interesting unit for forestry and agricultural surveying is made by Parrot. The Sequoia uses GPS to mark the locations of images that it takes through its multiple lenses. The images comprise visible light, two infrared (IR) bands, green, and red light. This unit can be attached to any drone but comes fitted standard to their Disco-Pro AG fixed-wing drone. There are two parts, one that affixes to the top of the drone to collect sunlight information and GPS tracking data, and the camera piece that is attached to the bottom of the drone. This and other similar units from other vendors will become more commonplace as the agriculture and forestry uses for drones are fully realized.

There are rental houses that rent out drones equipped with IR cameras. If your use of this technology is occasional, then it may be cheaper to go the rental route rather than to pay for changing technological solutions that will improve year to year.

Emerging Subjects for the Tech-Savvy Pilot

Specialty Software

In an earlier section, we touched on 3rd part software such as Litchi. As neat as this software may be, there are specialty packages that have been developed for specific niche markets. One such market is the agricultural community. There are a number of drones being marketed with specialized agricultural inspection section. This is one market in which conventional fixed-wing drones tend to have a power-to-time advantage over rotorcraft drones.

Of course, most readers have heard about Amazon's initiative to use drones for rural delivery of small packages. The FAA has not ruled yet on how drones work beyond line of sight, but that will happen soon. Meanwhile Amazon and other delivery companies are developing software to permit this use of drones. The medical community is another potential emerging market for this kind of drone usage for delivery of medicines to difficult-to-reach locations as well as to be able to speed critical ("stat") samples from a patient to a lab for testing.

Other initiatives include precision mapping for drone use and realtime drone tracking systems to permit flight controllers to be completely aware of where drones are flying within the NAS. The next few years will bring incredible improvements in these technologies.

An interesting realm of software is found in the BVLOS (Beyond Visual Line of Sight) universe. Several software packages have been in active development to provide instrumentation and warning as BVLOS drone operation is poised to become a reality. An example of this is PrecisionHawk's LATAS which tracks all aircraft within the potential Dronespace including other drones equipped with LATAS instrumentation. The system can alert Latas-equipped drone operators and eventually ATC systems of potential problems as aircraft and/or drones infringe on each other's space. LATAS uses a lightweight device affixed to the drone to track its position in space and to relay this data back to its servers.

In addition to the comprehensive warning system it provides, LATAS also tracks flights in a device-agnostic way. A LATAS user can file an expected operations flight plan and then compare the actual flight against that plan to determine if there were deviations from the plan. DJI and other manufacturers include electronic logging but their solution is proprietary and limited. Systems such as

LATAS which operate outside of the realm of drone manufacturers' software offerings provide a truly portable and useful logging solution.

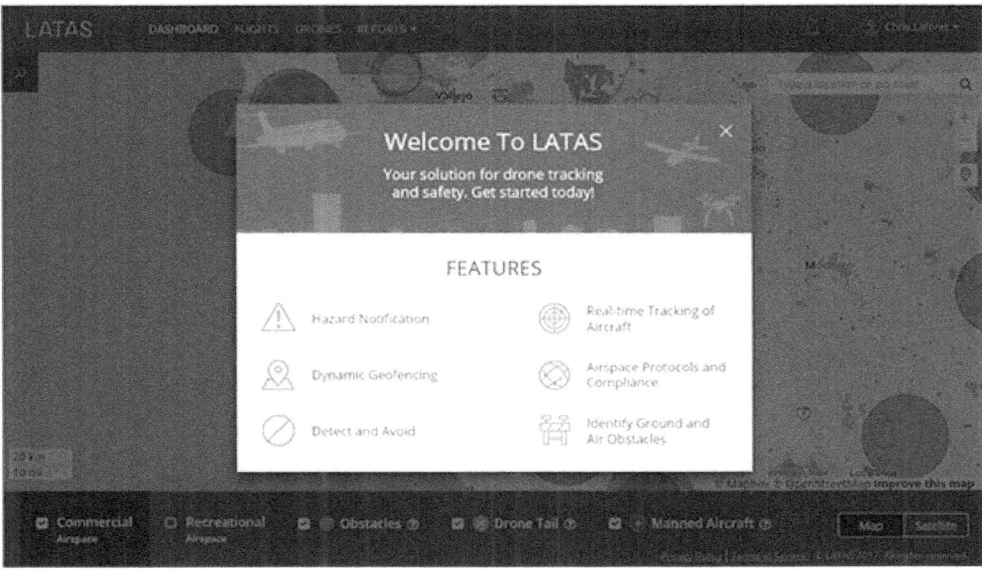

Figure 21 PrecisionHawk's LATAS Web Portal

Law Enforcement and Emergency Services

As their capabilities are becoming understood, drone operations are slowly gaining traction in the hands of the police and emergency services sectors. Programs for drone operations are being stood up in fire departments, rescue units, and police entities in order to provide tangible eye-in-the-sky support systems that can help ground operations.

Fire departments and rescue units have already demonstrated the power of being able to launch drones with IR (infrared) cameras to help direct searchers to missing or injured people. Some fire departments are experimenting with drones as aerial platforms to provide additional intelligence to their mobile command centers at large fires.

In September 2017, the North Carolina Highway Patrol authorized used of DJI Mavic drones equipped with pix4d software by their Collision Reconstruction Unit to map accidents when possible (see their official release at https://apps.ncdot.gov/newsreleases/details.aspx?r=14311). Tests proved that drone operations were significantly faster than the traditional methods and this could lead to roads being reopened quicker following a major accident.

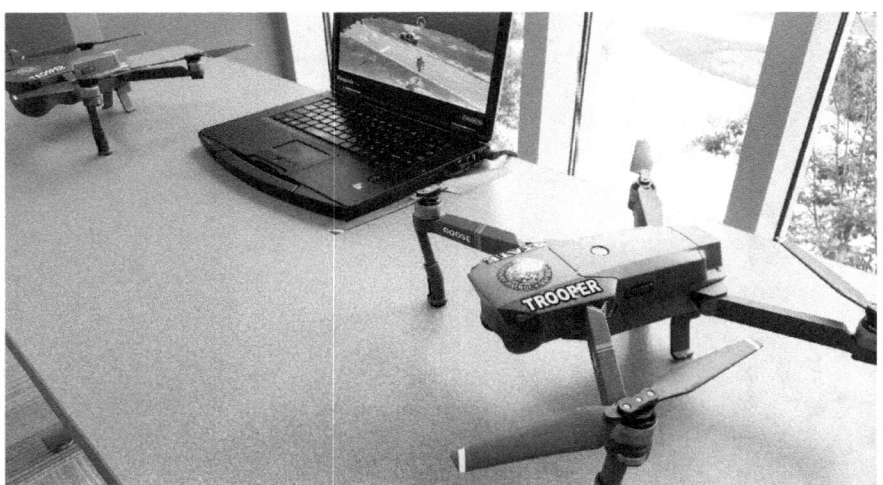

Figure 22 A pair of NC Highway Patrol Accident Reconstruction drones on display at an NCDOT conference in October 2017

Some Law Enforcement groups have begun using drones to provide intelligence on critical/dangerous warrant searches and while breaching buildings to make arrests. Of course, such surveillance use of drones is driven by warrants. Some groups have also entertained the ideas of weaponizing drones for Law Enforcement use, perhaps with non-lethal options. Maybe within a year or two, the first legitimate takedown of a criminal by a drone will become a reality.

Specialty Hardware

In addition to the special little attachments that are being created for drones, there are others that work on a much bigger level. An example of this is the company UAV Mobile Station, LLC (uavmobilestation.com) which has tackled the need for a safe, enclosed, connected, computing-intensive, and multi-pilot mobile office for drone teams to use while in the field. This solution will become even more effective as the BVLOS (Beyond Visual Line of Sight) regulations are unveiled in 2018 and later. Equipped with satellite-based internet, lighting, power solutions, antenna boosters, flat screens inside and outside the trailer, and a host of other communications perks, flight and search-and-rescue operations can be directed from inside or outside the trailer conveniently.

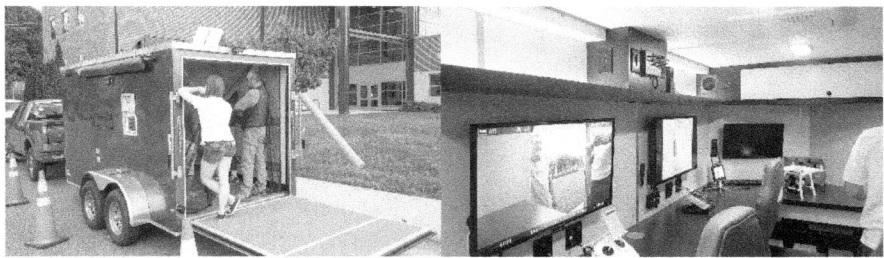

Figure 23 UAV Mobile Station trailer outside and inside shots at a demonstration site earlier in 2017

Some of the future of business in the Dronespace will certainly be centered on providing support for drone operations.

Directions the Dronespace is Heading

With the advent of concrete regulations concerning drone use there are No-Fly zones that are enforced. A number of drone software packages already have started incorporating No-Fly mapping for airports and governmental centers. These mappings range in activity from warning the operator, to requiring assent to override the No-Fly area if the operator has secured permission, to preventing the drone from entering or taking off within the zone.

Over time, it is expected that more and more mappings will be entered and instead of permitting an operator to merely enter a "Yes" to signify approval to fly in a No-Fly area, there may be some sort of coded key provided when permission is received by the operator from the controlling agency and which needs to be entered and approved by the drone software before flight will be permitted.

Another interesting direction that we are seeing the Dronespace evolving is how drone software tracks and logs flights on the manufacturer's corporate servers. DJI has this built into their Go App and others are certainly following suit. One use for this data is the manufacturers can gain real-life flight data that can be mined for improvements to their software. It also provides a benefit to the drone operator by keeping an automatic logbook but this data certainly will also be available for legal proceedings if a subpoena is served to the software manufacturer. It would make sense that this will become more iron-clad over time since regulatory agencies may require such data to prosecute felonious use of sUASes.

Drone Law

An interesting new arena is opening up concerning the laws surrounding drones. There are new laws appearing on the books concerning trespassing and stalking with drones. As drones become more ubiquitous, there will certainly be lawsuits generated by misuse of drones. There are already lawyers advertising that they are drone-related specialists. It will certainly be interesting to see where this goes in the future and serves to underline the need to be a respectable operator in the Dronespace.

During 2017, several interesting lawsuits against drone laws were prosecuted and won. One of the most talked-about one was the ruling in May by the DC Circuit Court of Appeals that struck down the FAA's requirement to register hobbyist drones. The suit had be brought by John Taylor against the FAA because the new requirement violated the FAA Reform and Modernization Act of 2012 which has provisions for hobbyist use of model aircraft.

An interesting case which will be replayed throughout the country is Singer v City of Newton (MA). At the end of 2016, the city passed a drone ordinance in which several provisions were overturned because the city attempted to potentially usurp the authority of the FAA, namely dealing with registration and airspace concerns.

Appendices

Checklists

The aviation world is filled with checklists. Every activity performed on an aircraft or with an aircraft has a checklist. Checklists exist for normal operation situations and others for emergency situations. To some, checklists may seem overkill but the fact remains that drilling checklists into one's being is good. Following the same steps each time ensures that nothing critical gets missed and this is especially good when an aircraft takes to the skies!

This section includes some of my personal checklists for my DJI Inspire 1 X5-RAW as samples that you might want to adapt to your own use. Some drones are more complex (e.g. DJI Matrice) and will require more extensive operations to become flight-ready while others may not need nearly the prep work in these checklists.

At the very least you should have a simple checklist in mind before you start a flying session. A basic list may be to check the props, turn on controller, turn on copter, check software versions for updates, calibrate compass, set home-point, and launch the craft.

Chris' DJI Inspire Preflight Checklist

Inspection of sUAS (before exiting travel mode)

1. Check prop blades for damage before they are mounted
2. Check optical sensors are clean
3. Check camera lens is mounted and confirm Lens Lock is clicked counterclockwise (LOCKED)
4. Check SD card is mounted in camera
5. Check SSD is mounted in camera
6. Check all parts to ensure none are loose
7. Check smart batteries for sUAS are charged
8. Check remote control station (CS) is properly charged
9. Check tablet for CS is properly charged

Preparation of sUAS for flight

1. Place sUAS on smooth surface

2. Insert smart battery into sUAS
3. Extend antennas on CS and ensure they are not crossed
4. Power on CS
5. Power on smart battery in sUAS
6. Confirm link between CS and sUAS indicates Master
7. Toggle landing gear lever on CS 4 times and confirm UAS extends from travel mode to flight mode
8. Power off smart battery in sUAS
9. Power off CS
10. Rotate the camera gimbal lock to unlocked position
11. Align white lines on gimbal connector and gimbal lock and insert gimbal connector
12. Rotate gimbal lock to LOCKED position and confirm LOCK
13. Rotate two keys on end of gimbal mounting plate 90 degrees into slots

14. Install propellers and ensure locks are clicked
15. Install tablet on CS holder(s) and connect USB

Preparation for Camera Operations

1. Filter or Balancing ring
   a. (EITHER) Does an ND filter need to be installed
   b. (OR) Does a Polarizer need to be installed
   c. (OR) Check camera has balancing ring installed
2. Check camera lens is set to AF mode
3. Check free space on SD card and clear/format if needed
4. Check free space on SSD and clear if needed
5. Check camera has lens hood installed

Chris' DJI Inspire Flight Operations Checklist

1. Ensure all crew have been briefed on the mission
2. Check radios if crew needs them for communication
3. Confirm there are no last minute adjustments to the mission and if so, communicate to crew

4. Ensure Preflight checklists for sUAS and Camera are complete
5. Set CS to P mode
6. Turn on CS
7. Confirm CS status LED is solid Red
8. Confirm Transformation Switch on CS for landing gear is down
9. Check for updates and install them to the CS or schedule for maintenance window
10. If using slave CS for camera
    a. On flight CS, set up Master under RC settings
    b. Power up camera CS
    c. On camera CS, select Slave under RC settings
    d. On camera CS, select Search for Master Controller
    e. Select Master Controller and enter password
    f. Confirm camera CS has status LED of Purple
11. Confirm Multimode is turned on in CS menu
12. Place sUAS on flat surface away from magnetic/electrical equipment or lines
13. Power on smart battery in sUAS and confirm Red/Green/Yellow then alternating Green/Yellow
14. Confirm (master) CS status LED is solid Green.
    a. (If using slave) Confirm (slave) CS status LED is solid Blue
15. Confirm DJI Go App on tablet is connected to sUAS
16. Note if there are updates for the sUAS and make a note to do them at later in Flight Log
17. If flying in a different location from previously, calibrate compass on sUAS
18. Wait for GPS signal to be acquired - signaled by Green flashes
19. (If flying GPS, turn of visioning positioning system to prevent issues with animals and sonar frequencies if needed)
20. Configure mission parameters into DJI Go App
21. Set up camera
    a. Check camera priority mode (A, S, M) and configure settings
    b. Check and set camera white balance (Sunny/Cloudy)
    c. Set still image recording format (Raw/JPEG)
    d. Set still image size (4:3/16:9)
    e. Set video recording format (RAW/H264)
    f. Set video recording size (4K/2.7K/1080) and frame-rate
    g. Set D-Log or None for color

22. Confirm gimbal is working properly (motion and level)
23. Mark flight start in Flight Log
24. Launch or Auto Launch (starts/takes off)
    a. Start engines and confirm proper operation
    b. Take off and hover at 2-6 feet and check controls are working as expected
25. Click the Transformation Switch on CS up to permit landing gear to raise

Chris' DJI Inspire Inside Flight Checklist

Additional items to check for Flight inside a Building

1. Step outside of the building and calibrate compass if not done already
2. Set return to home (RTH) altitude at 3 feet or something reasonable
3. Set loss-of-signal response to HOVER (or land) instead of RTH
4. Turn off GPS by moving control to A(tti)
5. Determine if to permit landing gear to go up or to remain up

Flight items to be aware of

1. Note that flying within several feet of a roof will create a suction effect
2. Note that flying over a pit will change the behavior of the sUAS
3. Note that flying near a wall (3 feet or so) will create a suction effect
4. Note that flying over objects will change the altitude of the sUAS unexpectedly

Chris' DJI Inspire Postflight Checklist

Shutdown of sUAS

6. Land the sUAS and shut down motors
7. Power off smart battery in sUAS
8. Record end of flight in Flight Log and make any notations
9. Inspect the sUAS and note in Flight Log if there are items that will require service
10. Exit DJI Go App and power off tablet
11. Power off Slave CS if in use

12. Either perform the following procedures (if there are no other flights) else use **Flight Operation step 12** checklist (if another flight is requires)
    a. Remove gimbal and camera from sUAS
    b. Place sUAS on smooth surface
    c. Power on smart battery in sUAS
    d. Toggle landing gear lever on CS 4 times and confirm UAS extends from flight mode to travel mode
    e. Power off smart battery from sUAS
    f. Remove spent battery from sUAS and place into case
    g. Remove props from sUAS
    h. Power down (master) CS
    i. Pack everything away

Frequently-Asked Questions

This section will deal with a series of questions that are commonly asked concerning drones. It is important to keep up with the current state of the law concerning sUAS registration and usage so these responses should serve as a springboard into doing further research. They may have been true at the time of publication but laws may have changed since then. It is incumbent on the reader to confirm that they are still correct.

1. **I have a new drone that weighs 0.5 pounds with the battery. I have a camera and gimbal that I purchased for it and bolted into place and that adds an additional 0.06 pounds to the drone. Do I need to register this drone with the FAA?**

   Yes, it needs to be registered after the modification. The rule is that if the drone weighs 0.55 pounds (8.8 ounces) or more at takeoff, it must be registered with the FAA before it is flown.

2. **Is it true that as a non-commercial hobbyist drone pilot, I am not ever permitted to fly within 5 miles of an airport?**

   No, that is not true. The FAA states concerning the regulations under the Special Rule for Model Aircraft that the operator is to "provide prior notification to the airport and air traffic control tower, if one is present, when flying within 5 miles of an airport." (www.faa.gov/uas/faqs) Before flying a drone in this radius, you must contact the airport or control tower for a towered airport and inform them where you will be flying (e.g. 3 miles Southwest of the airport), at what time you plan to fly and for how long you plan to fly, and how high (up to 400 ft AGL) your ceiling will be.

   Unfortunately, B4UFLY does not provide phone numbers for the airport but other resources such as FlightIntel or Avare on a smartphone or the online Chart Supplements Navaid link after a search on the FAA website at (www.faa.gov/air_traffic/flight_info/aeronav/digital_products/dafd/search/) can get you the phone number to the airport manager. In an airport controlled by a control tower, the manager can connect you to the tower or might take the information and relay it themselves. Control tower phone numbers in some cases may also be possibly found online.

Depending upon where you are planning to fly relative to the traffic pattern in place at the airport, you can be denied that flight. In their FAQs (see the link above), the FAA states "an airport operator can object to the proposed use of a model aircraft within five miles of an airport if the proposed activity would endanger the safety of the airspace. However, the airport operator cannot prohibit or prevent the model aircraft operator from operating within five miles of the airport." That objection, if it comes, should be considered a denial because continuing the plan to execute that flight can be used as "evidence that the operator was endangering the safety of the National Airspace System."

3. **Can a commercial drone operator be permitted to fly within the inner rings of class-B or class-C airports, within the 10 NM ring of a class-D, or within a surface class-E area?**

No, not without a waiver granting airspace authorization. This waiver can be requested at least 90 days in advance of the date of the flight through the FAA's online portal (see Filing for Waivers from the FAA).

4. **Do I need to apply for an N-tail-number for my commercial drone?**

No, this is not necessary in most cases. An N-number is a standard number that is assigned to regular aircraft. A commercial drone can be registered online and be assigned a standard drone registration number unless the sUAS weighs 55 pounds or more, or you need to use it outside of the US airspace, or the ownership of the sUAS is in a voting trust to meet the U.S. citizenship requirements, or the drone's title is held in a trust.

5. **Am I limited to flying 400 feet above where I launched my drone?**

No, do not look at the altitude limitation in terms of a flat pancake 400 feet above your launch site. The rule sets this limit to 400 feet AGL, that is, 400 feet Above Ground Level. Thus, this height will fluctuate relative to the floor over which the drone is flying.

If you are on the side of a mountain, as you extend your flight over an area where the mountainside is lower than your launch position, your maximum permitted altitude while over that area will be lower than it is over that launch position. Let's say that your launch point is at 1,000 feet, you can fly up to 1,400 feet over that point. If you fly outward from the mountain, say to where the "ground level" is 200 feet lower than your launch position, then your maximum altitude while flying in that area is 1,200 feet. The same logic is true if you fly so that the "ground level" is now 300 feet above your launch point. While hovering over that higher area, your maximum altitude is now 1,700 feet.

Here is where it actually gets interesting. The rule is that while you are within 400 feet of a structure, regardless of if that structure is a building, a tower, or a mountainside, your maximum altitude is 400 feet above the tallest point of that structure.

6. **What do I need to do if a police officer approaches me while flying a drone?**

First of all, bear in mind that not all law enforcement officials are aware of all of the laws surrounding sUAS use. Be patient and be a good advocate for drone-users everywhere. If your drone is still in the air, ask them if they need you to bring it in for a landing.

The first piece of documentation that you need to provide is a copy of your drone registration. When you first registered your drone, you were given the opportunity to print out the registration slip which should live in your drone's carrying case and/or your wallet or purse. If your drone is on the ground, you can then show them that the registration number is clearly marked on the drone itself.

If you are flying the drone commercially, then show them your FAA license. In either case, you may need to show them a form of photo ID such as a driver's license. In a state which has a separate permitting requirement (such as North Carolina's www.ncdot.gov/aviation/uas/), you may also need to provide a copy of the permit.

Beyond this, you will need to determine what specifically the officer needs. It may be a question of if you have permission to launch and recover your craft in private property. It may be an issue of someone concerned that you may be spying on them. Whatever the case, just maintain a dialog with the officer and as long as you were not flying recklessly or illegally, you should be ok.

7. **If I hit a bird while flying my drone, do I just ignore it?**

No, you should not if you are flying a drone commercially. It is preferable to file a report even if flying as a hobbyist since all bird-strike data is important for flight safety. This can be done using a form (FAA Form 5200) or online at http://wildlife.faa.gov/strikenew.aspx. You do not have to fear any sort of penalty from the FAA for the strike.

The best approach to avoid birds is to climb above them if you see them flying towards your drone.

8. **I want to fly my drone but the day is foggy. Is it ok to fly it?**

No, you cannot fly a drone legally in clouds. Fog and mist are essentially clouds that are near to or touching the ground. FAA rules stipulate that there must be a 3 statute mile or greater visibility to fly a drone legally and a drone is not permitted to fly above clouds, within 500 feet below a cloud (fog included), or 2,000 feet horizontally away from a cloud.

This is a rule that may be waived for a commercial drone operator using the FAA's waiver application program.

9. **How or when do I need to file a report if I crash my drone?**

A commercial drone PIC must report any accident of the sUAS to the FAA within a period of 10 days from the incident if either there has been any serious injury to a person that has led to them losing consciousness or has made them have to stay in a hospital, or there has been damage to any property other than the drone itself that has cost over $500 to repair it or replace it.

The report can be filed by phone with the FAA through an FAA Regional Operations Center or a Flight Standards District office or it can be filed online at www.faa.gov/uas/report_accident/.

There is the possibility if you are flying a commercial drone mission, if someone is killed or seriously injured or a serious incident such as a fire, flyaway, or collision with another aircraft, that a report may need to be filed with the NTSB (844-373-9922 or online at www.ntsb.gov/pages/report.aspx).

10. **What should I do if a low-flying helicopter or airplane flies towards the area my drone is flying in?**

See-and-avoid is the drone operator's mantra. If you or a VO spot an aircraft coming towards your sUAS operations area, your immediate response is to get the drone down below them.

If you cannot safely get below them, you are permitted to fly up higher than the aircraft even if you have to violate the 400 foot AGL rule. You are permitted to deviate from any part-107 rule to respond to an in-flight emergency. The PIC does not need to report the violation of the rule unless the FAA requests a written report.

Flying with pre-considered contingency plans is the best answer to this situation, should it arise. Know potential landing zones, or areas you can duck down between trees and be out of the main airspace, so if an aircraft comes at you out of thin air, you can get to one of these preplanned locations and be out of their way.

11. **Can I fly my drone at night?**

No. Drones are not permitted to fly in conditions other than civil twilight, about 30 minutes before sunrise and 30 minutes after sunset and they must have anti-collision lights that are visible for 3 miles. This is a rule that may be waived for a commercial drone operator using the FAA's waiver application program.

12. **Can anything happen to my commercial pilot license if I am convicted of driving under the influence or if I refuse a test when pulled over for suspicion of driving under the influence?**

    Yes. Refusing to take a test or failing a sobriety test requires the airman to submit a notification letter to the FAA. Failure to send this notification is grounds to have the license revoked or suspended, and applications for any rating, certificate, or authorization to be denied up to one year after the DUI/DWI action. The notification letter can trigger an investigation which may also result in the suspension or revocation of your FAA license.

13. **How can I locate other drone enthusiasts in my area?**

    I recommend using meetup.com to see if there are meetups near you. There are a large number of meetup groups scattered around the country and they offer programs or flying gatherings on a weekly or monthly basis.

    Also you can search Meetup and the Internet to find if there are any RC parks in your area. RC Airplane World offers lists of RC parks around the world at www.rc-airplane-world.com/rc-airplane-clubs.html. Most parks require an annual membership fee but there are many benefits to them. First, their location and operations are sanctioned by local airports and second, there are large number of seasoned RC and drone pilots among their membership.

14. **What should I do if someone asks, "Can I fly your drone?"**

    This depends on the situation and who is asking. In most cases, the answer is "No" but there can be some opportunities that arise from this question to advocate positively for drone use.

    The most common time I have heard this question is when kids come over and say, "That is so cool, can I try it?" The answer is definitely a "No" because the FAA tends to look at age 13 as a magic age. They don't state that someone younger cannot operate the controls but at 13 years old, an individual can register a drone with them. If someone younger than 13 wants to fly, it is up to their parents/guardians to accommodate this.

However, even though I may say no to them, I do try to explain to them that these big drones are not toys and anyone using the controls needs to have experience with them. Then I tell them to ask their folks for a small toy drone so they can start flying and then maybe one day get one of the bigger drones. Ultimately, the decision is up to you and is based on the relationship you may have with the young person who might be your niece or cousin versus some random child who just ambled over to you.

If you are flying a commercial operation, the answer is always an immediate "No." As a PIC, or a member of a crew under a PIC, the operator(s) of the drone and its mission are set. The drone in this case neither is a toy nor is it being used for fun.

However, if you are flying for fun, you may want to advocate for drones by explaining some basics of the controller (e.g. "Here is how you can make the drone climb or come down...") and without giving up the controller, permit someone to move the control and see what it does. Then you can show them another aspect of the controller, and let them try it. If your drone uses dual controllers, you might let them take the camera controller and try moving the gimbal around, click off some photos, or take a short video. Who knows, just that little bit of contact with a drone might bring someone else into the hobby.

15. How to deal with neighbors while flying a commercial mission?

It is important that any mission is flown over property that you are permitted to fly over. This permission may be explicit permission from the customer or may be implicit if part of the mission will traverse public space (such as roads, city or county parks, etc.). If a mission will require transit of someone else's property, it is always best to secure permission in advance. If you are denied, at least you will not discover this after the fact when a neighbor files a nuisance report against you.

In some states, flying over a person's property can be considered trespassing and if a camera is in use, may be considered peeping activity which is considered a sexual misdemeanor or felony, so it is even more important to be covered. It is best to secure such permission on a signed form. It doesn't have to be elaborate, just something that states that the undersigned person has permission to grant you their permission to overfly their land during a specified time range on a specific date.

Sometimes, neighbors may amble over during a shoot and ask what you are doing. Remember, drone advocacy comes first. Let them know you are shooting on this property with permission and answer their questions. They may just want to know more about this fascinating new technology, so encourage them to look at the screen and as much as you can, let them look on. Of course, you may need to ask them to stand in a different location so they won't show up on camera, but all but the most unreasonable person can understand this.

If they do become a problem, you may have to revert to letting the owner of the property deal with them or even involve the police. There are actually people who think that they have the right to shoot things out of the sky thanks to some bad press on the Internet. It is a Federal crime to shoot at a drone. The FAA treats this as no different from shooting at any other aircraft per 18 USC 32. While it may be a rare occurrence, do not hesitate to involve law enforcement if someone communicates threats or acts in any way that you or your crew interpret as them "going off the rails."

16. What drone should I buy my kid for a present?

As with all gifts, this depends upon their age and maturity. A young child should start with a simple $50 toy drone that is easy to repair and not a major loss if abandoned or mistreated. There are a number of decent brands mentioned in this book.

If a child is very mature, you may opt to get one of the larger drones for them. An entry-level larger drone may make a perfect starting point for them. I would advise you to take the time to explain the rules of the Dronespace and the dangers of misusing a drone. It would not hurt to get them involved in a model-aircraft park if you live near one.

A child who has proven themselves on a toy drone or an entry-level larger drone may be well qualified to move up a step to something more substantial and expensive. Use good parental judgment as you make this decision. There are many parents who purchase $1000+ first-time drones for their kids and end up with crashed and demolished pieces of plastic and metal with a few days of the gift being opened.

17. What are good practices for flying a drone in cold weather?

Batteries are adversely affected by cold so be very cautious with trusting them in flying far or high. At any moment, a cold battery can lose charge so be extra vigilant and even if your system meter indicates a full battery charge, consider that your flight time will be shorter than normal.

A great option is to get a small cooler or cooler bag, pop in some hand warmers to make it nice and inside, and keep your batteries in the cooler until you have to use them. Some drones (e.g. DJI Inspire) offer foam insulating stickers to attach to a battery before use in low temperatures. If the temperature is below the stated temperature, do use these to further conserve battery temperatures.

Finally, do not fly while it is snowing or sleeting. Most drones are not built to operate in moist conditions and thus can be damaged by melting precipitation. Of course, bear in mind that the FAA also prohibits flight when it is foggy (anything less than 3 statute miles of visibility horizontally) without a Part 107 waiver or a 333 exemption.

18. Can I fly my US-registered drone in a foreign country?

Yes, but you need to follow the rules for that country's use of airspace. It is best if you check online before you travel to determine what you will need to do when you arrive there. Some countries are very drone-friendly and others are drone-averse. Some require registration (like the US does) for visitor's drones. Hobby use rules may differ from commercial use rules as they do in the US also. Note that US-registered commercial drones that will be used overseas need to be registered by the FAA with an N-tail number (see question #4 above).

Of course, the benefit of having a drone with you on an overseas trip is that you can capture definitively different shots like this.

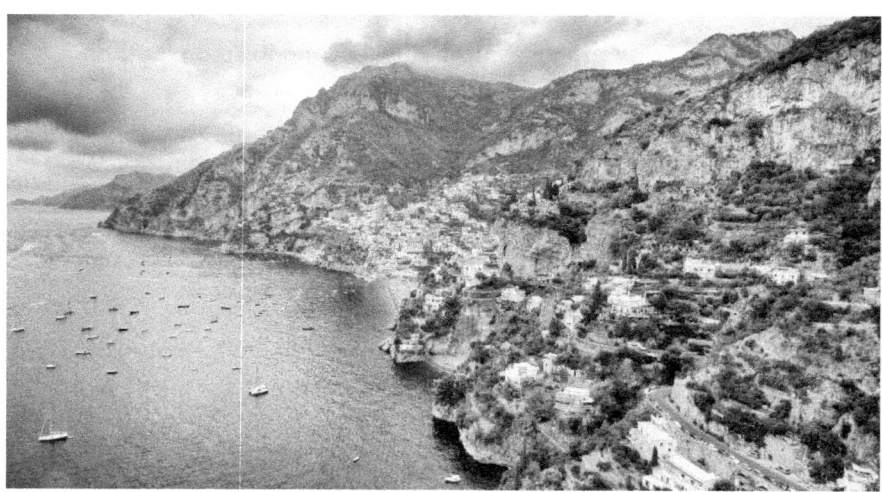

Figure 24 Positano, Italy. Taken with an Inspire 1 drone a quarter mile off of the Amalfi Coast over the Mediterranean Sea at an elevation of 250 feet. The drone was launched from Laurito Beach to the delight of onlookers. Italy proved to be reasonably drone-friendly. September 2016, Courtesy of Joe Valasquez and DroneScape LLC

Glossary of Terms and Acronyms

AGL	Above Ground Level, the altitude an object is above the ground directly beneath it. AGL is not an absolute altitude, unlike MSL.
ATC	Air Traffic Control, a system of ground-based controllers who direct traffic through controlled airspace and on the ground. ATC can also advise traffic that is in non-controlled airspace.
BVLOS	Beyond Visual Line of Sight, regulations that will permit drone operations to be directed using technologies other than a direct visual observation.
Controlled Airspace	Airspace within which air traffic control (ATC) services are provided to aircraft. Controlled airspace is defined as Class-A, Class-B, Class-C, Class-D, and Class-E. All drone flight is prohibited in this airspace without specific permission.
CTAF	Common Traffic Advisory Frequency, a VHF radio frequency used at non-towered airports that permit pilots to coordinate their taxiing, takeoff, and landing maneuvers with each other, and to provide position reports while entering, exiting, or participating in the airport's traffic pattern. CTAF does not require a ground station to be manned (MULTICOM) but it can be a UNICOM station.
Gimbal	A special device which rides on axles at right-angles to each other and which permits an instrument or camera to remain unaffected by movement in any of the directions of those axles. Gimbals can either be 2-D (only two planes) or full 3-D (all three planes).
Headless mode	See Intelligent Orientation Control

Intelligent Orientation Control	This is a mode for a drone controller in which no matter which way the drone is facing, the right hand stick convert forward/backward and left/right movements of the stick into commands to the drone that make it move further/closer and left/right relative to the person operating the controller. Forward does not have anything to do with where the nose of the drone is pointing.
IOC	See Intelligent Orientation Control
LAANC	Low-altitude Authorization and Notification Capability, a system that will enable certain drone software to dynamically request ad-hoc access to controlled airspace (Classes B, C, D, E) for Part 107 users.
METAR	A standardized aviation weather report emitted generally hourly from an airport weather station and which describes the observed conditions at that airport. METARs can also include information concerning changes to the weather. METARs are used in preflight planning to determine the meteorological conditions surrounding a flight.
MSL	Mean Sea Level, the altitude an object is above sea level. These altitudes are absolute, unlike AGL.
MULTICOM	A VHF frequency allocation used by non-towered airports or towered-airports during non-manned hours. A MULTICOM is not manned at the ground station. It permits CTAF operations.
Non-Controlled Airspace	Sometimes referred to as Uncontrolled Airspace. This is airspace in which air traffic control (ATC) services are either not practical to provide or are not deemed to be necessary. Class-G airspace is non-controlled and is open for drone flight without securing ATC permission.
NOTAM	See Notices to Airmen

Notices to Airmen	The FAA defines this as notices containing information (not known sufficiently in advance to publicize by other means) concerning the establishment, condition, or change in any component (facility, service, or procedure of , or hazard in the National Airspace System) the timely knowledge of which is essential to personnel concerned with flight operations.
Pan	Swiveling a camera around an axis that extends from top to bottom to point the lens more to the left or to the right. A yaw of an aircraft with a fixed camera is equal to a pan.
PIC	See Pilot in Charge
Part 107	The FAA rules that went into force on August 29th, 2016 and which include all pilot and operating rules. Officially it is known as Part 107 of the Federal Aviation Regulations or 14 CFR Part 107.
Pilot in Charge	
Pitch	The axis of an aircraft that permits raising or lowering the nose.
Roll	The axis of an aircraft extending from nose to tail that permits the aircraft to twist left or right relative around that axis.
TFR	Temporary Flight Restrictions filed with or by the FAA that disallow some or all flights within a specified area for a specified time.
UNICOM	A VHF radio facility that is not operated by air traffic control (ATC) that permits users of an uncontrolled airport to interchange advisory information. UNICOM generally is found and used at low-traffic, non-towered airports. UNICOM, by definition, requires the ground station to be manned. A UNICOM frequency could

revert to a MULTICOM frequency for CTAF when the airport facilities are closed.

UTM

Unmanned aerial vehicle Traffic Management, a system being developed by NASA for the FAA. This system is slated to be released in 2019 to enable drones to be used in autonomous applications (e.g. package delivery).

Visual Observer

A person who is tasked with tracking the location and movement of a drone and relaying that information back to the Pilot-in-Charge.

VO

See Visual Observer

Waypoint

A marked point in 3D space (normally in drone flight software)

Yaw

The axis of an aircraft which extends from top to bottom and allows the nose to swivel left or right.

(This page is left intentionally blank)

www.ingramcontent.com/pod-product-compliance
Lightning Source LLC
Chambersburg PA
CBHW072217170526
45158CB00002BA/638